Lecture Notes in Artificial Intelligence 6384

Edited by R. Goebel, J. Siekmann, and W. Wahlster

Subseries of Lecture Notes in Computer Science

Javier Larrosa   Barry O'Sullivan (Eds.)

Recent Advances in Constraints

14th Annual ERCIM International Workshop
on Constraint Solving
and Constraint Logic Programming, CSCLP 2009
Barcelona, Spain, June 15-17, 2009
Revised Selected Papers

 Springer

Series Editors

Randy Goebel, University of Alberta, Edmonton, Canada
Jörg Siekmann, University of Saarland, Saarbrücken, Germany
Wolfgang Wahlster, DFKI and University of Saarland, Saarbrücken, Germany

Volume Editors

Javier Larrosa
Technical University of Catalonia, Department of Software
C. Jordi Girona 1-3, 08034 Barcelona, Spain
E-mail: larrosa@lsi.upc.edu

Barry O'Sullivan
University College Cork, Cork Constraint Computation Centre
Western Road, Cork, Ireland
E-mail: b.osullivan@cs.ucc.ie

ISSN 0302-9743   e-ISSN 1611-3349
ISBN 978-3-642-19485-6   ISBN 978-3-642-19486-3 (eBook)
DOI 10.1007/978-3-642-19486-3
Springer Heidelberg Dordrecht London New York

Library of Congress Control Number: 2011921815

CR Subject Classification (1998): I.2.3, F.3.1-2, F.4.1, D.3.3, F.2.2, G.1.6, I.2.8

LNCS Sublibrary: SL 7 – Artificial Intelligence

Typesetting: Camera-ready by author, data conversion by Scientific Publishing Services, Chennai, India

Printed on acid-free paper

Springer is part of Springer Science+Business Media (www.springer.com)

Preface

This volume contains the selected technical papers from the 2009 ERCIM Workshop on Constraint Solving and Constraint Logic Programming held on June 15th–17th, 2009 at the Technical University of Catalonia (UPC) in Barcelona, Spain. This event was run on behalf of the ERCIM Working Group on Constraints[1]. ERCIM, the European Research Consortium for Informatics and Mathematics, aims to foster collaborative work within the European research community and to increase co-operation with European industry. Leading research institutes from 18 European countries are members of ERCIM. The ERCIM Constraints working group aims to bring together ERCIM researchers that are involved in research on the subject of constraint programming and related areas.

Constraints have recently emerged as a research area that combines researchers from a number of fields, including artificial intelligence, programming languages, symbolic computing, and computational logic. Constraint networks and constraint satisfaction problems have been studied in artificial intelligence since the 1970s. Systematic use of constraints in programming emerged in the 1980s. The constraint programming process involves the generation of requirements (constraints) and the solution of these requirements, by specialised constraint solvers. Constraint programming has been successfully applied in numerous domains. Recent applications include computer graphics (to express geometric coherence in the case of scene analysis), natural language processing (construction of efficient parsers), database systems (to ensure and/or restore consistency of the data), operations research problems (like optimization problems), molecular biology (DNA sequencing), business applications (option trading), electrical engineering (to locate faults), circuit design (to compute layouts), etc. Current research in this area deals with various foundational issues, with implementation aspects and with new applications of constraint programming. The concept of constraint solving forms the central aspect of this research.

The 2009 workshop programme comprised invited talks from Robert Nieuwenhuis (UPC, Spain) and Helmut Simonis (UCC, Ireland). The main technical programme also comprised talks from many constraints researchers on current aspects of their research agendas.

We would like to sincerely thank Dolors Padrós (UPC) for her assistance in preparing for this event, as well as Robert Nieuwenhuis and Helmut Simonis for their invited talks. We would also like to thank our sponsors who provided the financial support necessary to make this event a success. Finally, we would like to sincerely thank the authors of papers, the speakers, and the attendees, for such an interesting and engaging programme.

Javier Larrosa
Barry O'Sullivan

---

[1] http://wiki.ercim.org/wg/Constraints

Workshop Organization

CSCLP 2009 was organized by the ERCIM Working Group on Constraints.

Workshop Chairs

Javier Larrosa   Technical University of Catalonia, Spain
Barry O'Sullivan   University College Cork, Ireland

Additional Reviewers

Stefano Bistarelli   Ian Miguel
Hadrien Cambazard   Karen Petrie
François Fages   Luis Quesada
Ian Gent   Helmut Simonis
Emmanuel Hebrard   Armin Wolf
Alan Holland   Roie Zivan

Invited Speakers

Robert Nieuwenhuis   Technical University of Catalonia, Spain
Helmut Simonis   University College Cork, Ireland

Administrative Support

Dolors Padrós   Technical University of Catalonia, Spain

Sponsors

Association for Constraint Programming
Cork Constraint Computation Centre
ERCIM Working Group on Constraints
Technical University of Catalonia (UPC), Spain

Table of Contents

Solving Weighted Argumentation Frameworks with Soft Constraints*

Stefano Bistarelli[1,2], Daniele Pirolandi[1], and Francesco Santini[1,3]

[1] Dipartimento di Matematica e Informatica, Università di Perugia, Italy
{bista,pirolandi,francesco.santini}@dmi.unipg.it
[2] Istituto di Informatica e Telematica (CNR), Pisa, Italy
stefano.bistarelli@iit.cnr.it
[3] Dipartimento di Scienze Università "G. d'Annunzio", Pescara, Italy
santini@sci.unich.it

Abstract. We suggest soft constraints as a mean to parametrically represent and solve "weighted" Argumentation problems: different kinds of preference levels related to arguments, e.g. a score representing a "fuzziness", a "cost" or a probability level of each argument, can be represented by choosing different semiring algebraic structures. The novel idea is to provide a common computational and quantitative framework where the computation of the classical Dung's extensions, e.g. the admissible extension, has an associated score representing "how much good" the set is. Preference values associated to arguments are clearly more informative and can be used to prefer a given set of arguments over others with the same characteristics (e.g. admissibility). Moreover, we propose a mapping from weighted Argumentation Frameworks to *Soft Constraint Satisfaction Problems* (*SCSPs*); with this mapping we can compute Dung semantics (e.g. admissible and stable) by solving the related SCSP. To implement this mapping we use JaCoP, a Java constraint solver.

1 Introduction

Interactions are a core part of all multi-party systems (e.g. multi-agent systems). *Argumentation* [11] is based on the exchange and the evaluation of interacting arguments which may represent information of various kinds, especially beliefs or goals. Argumentation can be used for modeling some aspects of reasoning, decision making, and dialogue. For instance, when an agent has conflicting beliefs (viewed as arguments), a (nontrivial) set of plausible consequences can be derived through argumentation from the most acceptable arguments for the agent. Argumentation can be seen as the process emerging from exchanges of among agents to persuade each other and and bring about a change in intentions [20,18]. Argumentation has become an important subject of research in Artificial Intelligence and it is also of interest in several disciplines, such as Logic, Philosophy and Communication Theory [22].

---

* Research partially supported by MIUR PRIN 20089M932N project: "Innovative and multi-disciplinary approaches for constraint and preference reasoning".

J. Larrosa and B. O'Sullivan (Eds.): CSCLP 2009, LNAI 6384, pp. 1–18, 2011.
© Springer-Verlag Berlin Heidelberg 2011

Many theoretical and practical developments build on Dung's seminal theory of argumentation. A *Dung Argumentation Framework* (*AF*) is a directed graph consisting of a set of arguments and a binary conflict based *attack relation* among them. The sets of arguments to be considered are then defined under different semantics, where the choice of semantics equates with varying degrees of scepticism or credulousness. Solutions are defined through application of an ""acceptability" calculus", whereby an argument $A \in Args$ is said to be acceptable with respect to $S \subseteq Args$ iff any argument B that attacks A is itself attacked by some argument C in S (any such C is said to reinstate A). inclusion) set such that all its contained arguments are acceptable with respect to S, then S is said to be an extension under the *preferred semantics*. any theory of argumentation is the selection of acceptable sets of arguments, based on the way arguments interact. Intuitively, an acceptable set of arguments must be in some sense coherent and strong enough (e.g. able to defend itself against all attacking arguments).

The other ingredient in our research is Constraint Programming [23], which is a powerful paradigm for solving combinatorial search problems that draws on a wide range of techniques from artificial intelligence, computer science, databases, programming languages, and operations research. The idea of the semiring-based formalism [7,5] was to further extend the classical constraint notion by adding the concept of a structure representing the levels of satisfiability of the constraints. Such a structure (see Sec. 3 for further details) is a set with two operations: one + is used to generate an ordering over the preference levels, while × is used to combine these levels. Because of the properties required on such operations, this structure is similar to a semiring (see Sec. 3). Problems defined according to this semiring-based framework are called *Soft Constraint Satisfaction Problems* (SCSPs).

In this paper we show that different weighted AFs based on fuzziness, probability or a preference in general (and already studied in literature, e.g. in [22,3]), can be modeled and solved with the same soft constraint framework by only changing the related semiring in order to optimize the different criteria. Also classical AFs can be represented inside the soft framework by adopting the *Boolean* semiring. We provide a mapping from AFs to (S)CSPs in a way that the solution of the SCSP consists in the "best" desired extension, where "best" is computed by aggregating (with ×) the preference scores of all the chosen arguments, and comparing the final values (with +). The classical extensions of Dung can be found with our mapping, i.e. admissible, preferred, complete, stable and grounded ones. At last, we show an implementation of a CSP with *JaCoP* [21], a Java Constraint Programming solver.

Clearly, the classical attack relationship is not enough informative to deal with problems where we however need to take a decision: suppose a judge must decide between the arguments of two parties, and often no conclusive demonstration of the rightness of one side is possible. The arguments will not have equal value for the judge and the case will be decided by the judge preferring one argument over the other [22]. Moreover, having a quantitative framework permits us to quantify the aggregation of chosen arguments and to prefer a set of arguments over another. Examples in the real world are represented by scores given to comments in *Youtube* or news in *Slashdot*, or topics in Discussion Fora in general [17]. As the set of arguments gets wider, the search of the best solutions becomes a demanding task, and constraint-based frameworks come with many and powerful solving techniques: notice that deciding if a set is a preferred extension is a *CO-NP*-complete problem [4]. Moreover, preference score can be used to cut not promising solutions during the search and, however, to refine it by finding the only the best solutions. In this paper we start from qualitative argumentation [22,3,2] and we move towards a quantitative solution.

Notice that our soft constraint framework is able to solve all crisp and weighted extensions of Dung shown in Sec. 6 in a parametric way; therefore, the strength of this paper is to propose a general way to solve AFs.

The remainder of this paper is organized as follows. In Sec. 2 we report the theory behind Dung Argumentation, while in Sec. 3 we summarize the background about soft constraints. Section 4 shows the basic idea of weighted AF based on semirings; in Sec. 5 we propose the mapping from AFs to SCSPs, the proofs of their solution equivalence and we show a practical encoding in JaCoP. A comparison with related work is given in Sec. 6. Finally, Sec. 7 presents our conclusions.

2   Dung Argumentation

In [11], the author has proposed an abstract framework for argumentation in which he focuses on the definition of the status of arguments. For that purpose, it can be assumed that a set of arguments is given, as well as the different conflicts among them. An argument is an abstract entity whose role is solely determined by its relations to other arguments.

Definition 1. *An Argumentation Framework (AF) is a pair $\langle \mathcal{A}_{rgs}, R \rangle$ of a set $\mathcal{A}_{rgs}$ of arguments and a binary relation R on $\mathcal{A}_{rgs}$ called the attack relation. $\forall a_i, a_j \in \mathcal{A}_{rgs}, a_i R a_j$ means that a_i attacks a_j. An AF may be represented by a directed graph (the interaction graph) whose nodes are arguments and edges represent the attack relation. A set of arguments $\mathcal{B}$ attacks an argument a if a is attacked by an argument of $\mathcal{B}$. A set of arguments $\mathcal{B}$ attacks a set of arguments $\mathcal{C}$ if there is an argument $b \in \mathcal{B}$ which attacks an argument $c \in \mathcal{C}$.*

The "acceptability" of an argument [11] depends on its membership to some sets, called *extensions*. These extensions characterize collective "acceptability". Let $AF = \langle \mathcal{A}_{rgs}, R \rangle, \mathcal{B} \subseteq \mathcal{A}_{rgs}$.

In Fig. 1 we show an example of AF represented as an *interaction graph*: the nodes represent the arguments and the directed arrow from c to d represents the attack of c towards d, that is $c R d$. Dung [11] gave several semantics to "acceptability". These various semantics produce none, one or several acceptable sets of arguments, called extensions. In Def. 2 we define the concepts of conflict-free and stable extensions:

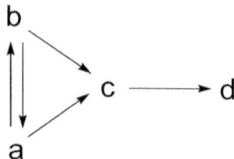

Fig. 1. An example of Dung Argumentation Framework; e.g. c attacks d

Definition 2. *A set $\mathcal{B} \subseteq \mathcal{A}_{rgs}$ is conflict-free iff it does not exist two arguments a and b in $\mathcal{B}$ such that a attacks b. A conflict-free set $\mathcal{B} \subseteq \mathcal{A}_{rgs}$ is a stable extension iff for each argument which is not in $\mathcal{B}$, there exists an argument in $\mathcal{B}$ that attacks it.*

The other semantics for "acceptability" rely upon the concept of defense:

Definition 3. *An argument b is defended by a set $\mathcal{B} \subseteq \mathcal{A}_{rgs}$ (or $\mathcal{B}$ defends b) iff for any argument $a \in \mathcal{A}_{rgs}$, if a attacks b then $\mathcal{B}$ attacks a.*

An admissible set of arguments according to Dung must be a conflict-free set which defends all its elements. Formally:

Definition 4. *A conflict-free set $\mathcal{B} \subseteq \mathcal{A}_{rgs}$ is admissible iff each argument in $\mathcal{B}$ is defended by $\mathcal{B}$.*

Besides the stable semantics, three semantics refining admissibility have been introduced by Dung [11]:

Definition 5. *A preferred extension is a maximal (w.r.t. cardinality) admissible subset of $\mathcal{A}_{rgs}$. An admissible $\mathcal{B} \subseteq \mathcal{A}_{rgs}$ is a complete extension iff each argument which is defended by $\mathcal{B}$ is in $\mathcal{B}$. The least (w.r.t. cardinality) complete extension is the grounded extension.*

A stable extension is also a preferred extension and a preferred extension is also a complete extension. Stable, preferred and complete semantics admit multiple extensions whereas the grounded semantics ascribes a single extension to a given argument system.

Notice that deciding if a set is a stable extension or an admissible set can be computed in polynomial time, but deciding if a set is a preferred extension is a $CO\text{-}NP$-complete problem [4].

3   Soft Constraints

A c-semiring [7,5] S (or simply semiring in the following) is a tuple $\langle A, +, \times, \mathbf{0}, \mathbf{1} \rangle$ where A is a set with two special elements $(\mathbf{0}, \mathbf{1} \in A)$ and with two operations $+$ and $\times$ that satisfy certain properties: $+$ is defined over (possibly infinite) sets of elements of A and thus is commutative, associative, idempotent, it is closed and $\mathbf{0}$ is its unit element and $\mathbf{1}$ is its absorbing element; $\times$ is closed, associative, commutative, distributes over $+$, $\mathbf{1}$ is its unit element, and $\mathbf{0}$ is its absorbing element (for the exhaustive definition, please refer to [7]). The $+$ operation defines a partial order $\leq_S$ over A such that $a \leq_S b$ iff $a + b = b$; we say that $a \leq_S b$ if b represents a value *better* than a. Other properties related to the two operations are that $+$ and $\times$ are monotone on $\leq_S$, $\mathbf{0}$ is its minimum and $\mathbf{1}$ its maximum, $\langle A, \leq_S \rangle$ is a complete lattice and $+$ is its lub. Finally, if $\times$ is idempotent, then $+$ distributes over $\times$, $\langle A, \leq_S \rangle$ is a complete distributive lattice and $\times$ its glb.

A *soft constraint* [7,5] may be seen as a constraint where each instantiation of its variables has an associated preference. Given $S = \langle A, +, \times, \mathbf{0}, \mathbf{1} \rangle$ and an ordered set of variables V over a finite domain D, a soft constraint is a function which, given an assignment $\eta : V \to D$ of the variables, returns a value of the semiring. Using this notation $C = \eta \to A$ is the set of all possible constraints that can be built starting from S, D and V. Any function in C involves all the variables in V, but we impose that it depends on the assignment of only a finite subset of them. So, for instance, a binary constraint $c_{x,y}$ over variables x and y, is a function $c_{x,y} : V \to D \to A$, but it depends only on the assignment of variables $\{x, y\} \subseteq V$ (the *support* of the constraint, or *scope*). Note that $c\eta[v := d_1]$ means $c\eta'$ where η' is η modified with the assignment $v := d_1$. Note also that $c\eta$ is the application of a constraint function $c : V \to D \to A$ to a function $\eta : V \to D$; what we obtain, is a semiring value $c\eta = a$. $\bar{\mathbf{0}}$ and $\bar{\mathbf{1}}$ respectively represent the constraint functions associating $\mathbf{0}$ and $\mathbf{1}$ to all assignments of domain values (i.e. the $\bar{a}$ function returns the semiring value a).

Given the set C, the combination function $\otimes : C \times C \to C$ is defined as $(c_1 \otimes c_2)\eta = c_1\eta \times c_2\eta$ (see also [7,5]). Informally, performing the $\otimes$ or between two constraints means building a new constraint whose support involves all the variables of the original ones, and which associates with each tuple of domain values for such variables a semiring element which is obtained by multiplying the elements associated by the original constraints to the appropriate sub-tuples.

Given a constraint $c \in C$ and a variable $v \in V$, the *projection* [7,5,6] of c over $V - \{v\}$, written $c \Downarrow_{(V \setminus \{v\})}$ is the constraint c' such that $c'\eta = \sum_{d \in D} c\eta[v := d]$. Informally, projecting means eliminating some variables from the support.

A SCSP [5] defined as $P = \langle C, con \rangle$ (C is the set of constraints and $con \subseteq V$, i.e. a subset the problem variables). A problem P is α-consistent if $blevel(P) = \alpha$ [5]; P is instead simply "consistent" iff there exists $\alpha >_S \mathbf{0}$ such that P is α-consistent [5]. P is inconsistent if it is not consistent. The *best level of consistency* notion defined as $blevel(P) = Sol(P) \Downarrow_{\emptyset}$, where $Sol(P) = (\bigotimes C) \Downarrow_{con}$ [5].

A SCSP Example. Figure 2 shows a weighted CSP as a graph: the semiring used for this problem is the *Weighted* semiring, i.e. $\langle \mathbb{R}^+, \min, \hat{+}, \infty, 0 \rangle$ ($\hat{+}$ is the arithmetic plus operation). Variables and constraints are represented respectively by nodes and by undirected arcs (unary for c_1 and c_3, and binary for c_2), and semiring values are written to the right of each tuple. The variables of interest (that is the set con) are represented with a double circle (i.e. variable X). Here we assume that the domain of the variables contains only elements a and b.

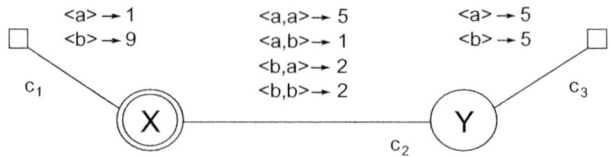

Fig. 2. A soft CSP based on a Weighted semiring

For example, the solution of the weighted CSP of Fig. 2 associates a semiring element to every domain value of variable X. Such an element is obtained by first combining all the constraints together. For instance, for the tuple $\langle a, a \rangle$ (that is, $X = Y = a$), we have to compute the sum of 1 (which is the value assigned to $X = a$ in constraint c_1), 5 (which is the value assigned to $\langle X = a, Y = a \rangle$ in c_2) and 5 (which is the value for $Y = a$ in c_3). Hence, the resulting value for this tuple is 11. We can do the same work for tuple $\langle a, b \rangle \rightarrow 7$, $\langle b, a \rangle \rightarrow 16$ and $\langle b, b \rangle \rightarrow 16$. The obtained tuples are then projected over variable x, obtaining the solution $\langle a \rangle \rightarrow 7$ and $\langle b \rangle \rightarrow 16$. The *blevel* for the example in Fig. 2 is 7 (related to the solution $X = a, Y = b$).

4   Weighted Argumentation

Weighted argumentation systems [9,12] extend Dung-style abstract argumentation systems by adding numeric weights to every node (or attack) in the attack graph, intuitively corresponding to the strength of the attack, or equivalently, how reluctant we would be to disregard it. To illustrate the need to extend the classical AF with preferences, we consider two individuals P and Q exchanging arguments A and B about the weather forecast (the example is taken from [22]):

P: Today will be dry in London since BBC forecast sunshine $= A$
Q: Today will be wet in London since CNN forecast rain $= B$

A and B claim contradictory conclusions and so attack each other. Under Dung's preferred semantics, there are two different admissible extensions represented by the sets $\{A\}$ and $\{B\}$, but neither argument is sceptically justified. One solution is to provide some means for preferring one argument to another in order to find a more informative answer, for example, the most trustworthy extension. For example, one might reason that A is preferred to B because the *BBC* are deemed more trustworthy than *CNN*. Suppose to have a fuzzy trust score associated with each argument, as shown in Fig. 3. This score, (between 0 and 1 that is between low and high trustworthiness) can be then used to prefer $\{A\}$ with a score of 0.9 over $\{B\}$ with a score of 0.7, i.e. forecast from *BBC* than from *CCN*.

In some works [16] the preference score is associated with the attack relationship instead of with the argument itself and, thus, it models the "strength" of the attack, e.g. a fuzzy attack. This model can be cast in ours by composing these strengths in a value representing the preference of the argument, as in

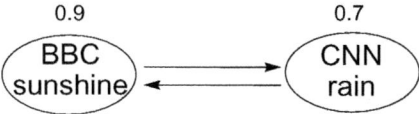

Fig. 3. The *CNN/BBC* example with trust scores

Fig. 4, where the trustworthiness of argument *CNN-rain* can be computed as
the mathematical mean (or in general a function ∘, as defined also in [8] for
computing the trust of a group of individuals) of the values associated with the
attack towards it, i.e. $(0.9 + 0.5)\backslash 2 = 0.7$. Computing a trust evaluation of a
node by considering a function of the links ending in it is a well-known solution,
e.g. the *PageRank* of *Google* [17]. By composing attack and support values, it is
also possible to quantitatively study bipolar argumentation frameworks [1].

Fig. 4. A fuzzy Argumentation Framework with fuzzy scores modeling the attack
strength

Notice that in [22,3,2] the preference among arguments is given in a qualitative way, that is argument *a* is better than argument *b*, which is better than
argument *c*; in this section we study the problem from a quantitative point of
view, with scores associated with arguments. We suggest the algebraic semiring
structure (see Sec. 3) as a mean to parametrically represent and solve all the
"weighted" AFs presented in literature (see Sec. 6), i.e. to represent the scores;
in the following we provide some examples on how semirings fulfil these different
tasks.

An argument can be seen as an event that makes the hypothesis true. The
credibility of a hypothesis can then be measured by the total probability that
it is supported by arguments. The proper semiring to solve this problem con-
sists in the *Probabilistic* semiring [5]: $\langle [0..1], max, \hat{\times}, 0, 1 \rangle$, where the arithmetic
multiplication (i.e. $\hat{\times}$) is used to compose the probability values together.

The Fuzzy Argumentation [26] approach enriches the expressive power of the
classical argumentation model by allowing to represent the relative strength
of the attack relationships between arguments, as well as the degree to which
arguments are accepted. In this case, the *Fuzzy* semiring $\langle [0..1], min, max, 0, 1 \rangle$
can be used.

In addition, the *Weighted* semiring $\langle \mathbb{R}^+, min, \hat{+}, 0, 1 \rangle$, where $\hat{+}$ is the arithmetic plus, can model the (e.g. money) cost of the attack: for example, during
an electoral campaign, a candidate could be interested in how many efforts or
resources he should spend to counteract an argument of the opposing party.

At last, with the *Boolean* semiring $\langle\{true, false\}, \vee, \wedge, false, true\rangle$ we can cast the classic AFs originally defined by Dung [11] in the same semiring-based framework.

Moreover, notice that the cartesian product of two semirings is still a semiring [7,5], and this can be fruitfully used to describe multi-criteria constraint satisfaction and optimization problems. For example, we can have both a probability and a fuzzy score given by a couple $\langle t, f \rangle$; we can optimize both costs at the same time.

We can extend the definitions provided in Sec. 3 in order to express all these weights of the attack relations with a semiring based environment. The following definitions model the semiring-based problem.

Definition 6. *A semiring-based Argumentation Framework (AF_S) is a quadruple $\langle \mathcal{A}_{rgs}, R, W, S\rangle$ of a semiring $S = \langle A, +, \times, \mathbf{0}, \mathbf{1}\rangle$, a set $\mathcal{A}_{rgs}$ of arguments, the attack binary relation R on $\mathcal{A}_{rgs}$, and a unary function $W : \mathcal{A}_{rgs} \longrightarrow A$ called the* weight function. *$\forall a \in \mathcal{A}_{rgs}$, $W(a) = s$ means that a has a preference level $s \in A$.*

Therefore, the weight function W associates each argument with a semiring value ($s \in A$) that represents the preference expressed for that argument in terms of cost, fuzziness and so on. For example, using the *Fuzzy* semiring $\langle[0..1], min, max, 0, 1\rangle$ semiring for the problem represented in Fig. 3 allows us to state that the admissible extension $\{A\}$ (with a score of 0.9) is better than the other admissible extension $\{B\}$ (with a s.core of 0.7) since $0.9 > 0.7$. Therefore, with an AF_S our goal is to find the extensions proposed by Dung (e.g. the admissible extensions), but with an associated preference value. Therefore, soft constraints can be used to solve these problems while considering also the best solution(s) (according to the notion of blevel, and to cut the solutions with a preference below a threshold α.

Example 1. Concerning the interaction graph in Fig. 5, it represents the Weighted AF_S $W = (\mathcal{A}_{rgs}, R)$ with $S = \langle\mathbb{R}^+, min, \hat{+}, \infty, 0\rangle$ and $\mathcal{A}_{rgs} = \{a, b, c, d, e\}$, $R(a, b), R(c, b), R(c, d), R(d, c), R(d, e), R(e, e)$ and $W(a) = 7, W(b) = 20, W(c) = 6, W(d) = 10, W(e) = 12$. Notice that e attacks itself, that is in contrast with itself, e.g. "We have sunshine and it's raining" (it may be possible).

5   Mapping AFs to SCSPs

Our second result is a mapping from AF (and AF_S) to (S)CSPs. Given an $AF_S = \langle \mathcal{A}_{rgs}, R, W, S \rangle$, we define a variable for each argument $a_i \in \mathcal{A}_{rgs}$, i.e. $V = \{a_1, a_2, \ldots, a_n\}$ and each of these argument can be taken or not, i.e. the domain of each variable is $D = \{1, 0\}$, and if it is taken, a cost in the semiring can be assigned, mapping the level of preference of this argument.

To represent the quantitative preference over arguments, in this mapping we need only unary soft constraints on each variable, while the other constraints modeling, for example, the conflict-free relationship (see Sec. 2) are crisp even

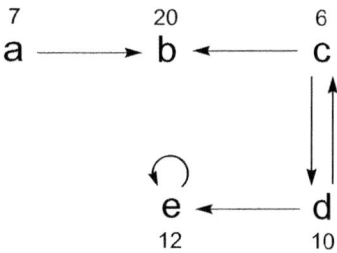

Fig. 5. An example of a weighted interaction graph

if represented in the soft framework. We plan to extend also these constraints to properly-said soft ones as suggested in Sec. 7. In the following explanation, notice that b attacks a meas that b is a parent of a in the interaction graph, and c attacks b attacks a means that c is a grandparent of a. To compute the (weighted) extensions of Dung we need to define specific sets of constraints:

1. **Preference constraints.** The weight function $W(a_i) = s$ $(s \in A)$ of an AF_S can be modeled with the unary constraints $c_{a_i}(a_i = 1) = s$, otherwise, when a_i is assigned to 0), the argument is not taken in the considered extension an so its cost must not be computed.

2. **Conflict-free constraints.** Since we want to find the conflict-free sets, if $R(a_i, a_j)$ is in the graph we need to prevent the solution to include both a_i and a_j in the considered extension: $c_{a_i,a_j}(a_i = 1, a_j = 1) = \mathbf{0}$. For the other possible assignment of the variables $((a = 0, b = 1)(a = 1, b = 0)$ and $(a = 0, b = 0))$, $c_{a_i,a_j} = \mathbf{1}$, since these assignments are permitted: in these cases we are choosing only one argument between the two (or none of the two) and thus, we have no conflict.

3. **Admissible constraints.** For the admissibility, we need that, if child argument a_i has a parent node a_f but a_i has no grandparent node a_g (parent of a_f), then we must avoid to take a_i in the extension because it is attacked and cannot be defended by any ancestor: expressed with a unary constraint, $c_{a_i}(a_i = 1) = \mathbf{0}$.

   Moreover, if a_i has several grandparents $a_{g1}, a_{g2}, \ldots, a_{gk}$ and only one parents a_f (child of $a_{g1}, a_{g2}, \ldots, a_{gk}$), we need to add a $k + 1$-ary constraint $c_{a_i,a_{g1},\ldots,a_{gk}}(a_i = 1, a_{g1} = 0, \ldots, a_{gk} = 0) = \mathbf{0}$. The explanation is that at least a a grandparent must be taken in the admissible set, in order to defend a_i from one of his parents a_f. Notice that, if a node is not attacked (i.e. he has no parents), he can be taken or not in the admissible set.

4. **Complete constraints.** To compute a complete extension $\mathcal{B}$, we need that each argument a_i which is defended by $\mathcal{B}$ is in $\mathcal{B}$ (see Sec. 2). This can be enforced by imposing that for each a_i taken in the extension, also all its $a_{s1}, a_{s2}, \ldots, a_{sk}$ grandchildren must be taken in the extension, i.e. $c_{a_i,a_{s1},\ldots,a_{sk}}(a_i = 1, a_{s1} = 1, \ldots, a_{sk} = 1) = \mathbf{1}$ and also if $a_i = 0$ this constraint is satisfied; $\mathbf{0}$ otherwise.

5. **Stable constraints.** If we have a child node a_i with multiple parents $a_{f1}, a_{f2}, \ldots, a_{fk}$, we need to add the constraint $c_{a_i, a_{f1}, \ldots, a_{fk}}(a_i = 0, a_{f1} = 0, \ldots, a_{fk} = 0) = \mathbf{0}$. In words, if a node is not taken in the extension (i.e. $a_i = 0$), then it must be attacked by at least one of the taken nodes, that is at least a parent of a_i needs to be taken in the stable extension (that is, $a_{fj} = 1$).

Moreover, if a node a_i has no parent in the graph, it has to be included in the stable extension (notice a_i cannot be attacked by nodes inside the extension, since he has no parent). The corresponding unary constraint is $c_{a_i}(a_i = 0) = \mathbf{0}$.

Notice that by using the *Boolean* semiring, also the class of preference constraints becomes crisp and we can consequently model classical Dung AFs, that is not weighted frameworks. The following proposition states the equivalence between solving an AF_S and its related SCSP.

Proposition 1 (Solution equivalence). *Given an $AF_S = \langle \mathcal{A}_{rgs}, R, W, S \rangle$ and $S = \langle A, +, \times, \mathbf{0}, \mathbf{1} \rangle$, the solutions of the related SCSP obtained with the mapping corresponds to find over AF_S the best(according to $+$)*

- *conflict-fee extensions by using preference and conflict-free constraint classes.*
- *admissible extensions by using preference, conflict-free and admissible constraint classes.*
- *complete extensions by using preference, conflict-free and complete constraint classes.*
- *stable extensions by using preference, conflict and stable constraint classes.*

By using the Boolean *semiring the solutions of the (S)CSP respectively correspond to all the classical admissible, complete and stable extensions of Dung [11].*

Moreover, to find the preferred extension (see Sec. 2) we simply need to find all the maximal (w.r.t. set inclusion) admissible extensions of $\mathcal{A}_{rgs}$, that is to find all the admissible sets (using the first three classes of constraints) and then returning only those subsets with the highest number of variables assigned to 1. Similar considerations hold for the grounded extension (see Sec. 2), that is we need to find all the complete extensions (the first four classes of constraints) and then to return only those subsets with the lowest number of variables assigned to 1^1.

As suggested in Sec. 4, an AF_S can be represented as a weighted interaction graph as in Fig. 5, where we instead suppose to use a *Weighted* semiring, i.e. $\langle \mathbb{R}^+, \min, \hat{+}, \infty, 0 \rangle$, e.g. the argument a has received 7 negative comments. The goal in this case is to choose the extensions of Dung and to minimize the sum of the negative comments at the same time.

Notice that the presented soft constraint framework can be easily used to solve argumentation problems with additional constraints, as proposed in [10] only for

---

^{1} Different interpretations of grounded/preferred extensions can be given by considering their cost instead of their the cardinality.

boolean constraints. We can find further requirements on the sets of arguments which are expected as extensions, like "extensions must contain argument a when they contain b" or "extensions must not contain one of c or d when they contain a but do not contain b".

Solving with JaCoP

The Java Constraint Programming solver [21], JaCoP in short, is a Java library, which provides Java user with Finite Domain Constraint Programming paradigm. It provides different type of constraints: most commonly used primitive constraints, such as arithmetical constraints, equalities and inequalities, logical, reified and conditional constraints, combinatorial (global) constraints. The last version of JaCoP proposes many features, such as pruning events, multiple constraint queues, special data structures to handle efficiently backtracking, iterative constraint processing, and many more [21]. Moreover, it can run also large examples, e.g. ca. 180000 constraints.

In Fig. 6 we show the definition in JaCoP of all the conflict-free and stable constraints used to solve the AF_S example in Fig. 5. The full description of the code can be found in Appendix A. Considering for example the first conflict-free constraint in Fig. 5, $v[0], v[1]$, means that the constraint is between a and b and $(1, 1)$ that the the constraint is not satisfied if both variables are taken in the set.

Considering the example in Fig. 5 the admissible sets are: $\{a\}, \{c\}, \{d\}, \{a, c\}, \{a, d\}$. Dung's semantics induce the following acceptable sets: one stable extension $\{a, d\}$, two preferred extensions $PE_1 = \{a, c\}, PE_2 = \{a, d\}$, three complete extensions $CE_1 = \{a, c\}, CE_2 = \{a, d\}, CE_3 = \{a\}$ and the grounded extension $\equiv \{a\}$. With our quantitative interpretation of AFs with preferences and considering the *Fuzzy* semiring $\langle \mathbb{R}^+, \min, \hat{+}, \infty, 0 \rangle$, we can prefer PE_1 over PE_2 $(W(a)\hat{+}W(c)) = 13$, $W(a)\hat{+}W(d) = 17$ and CE_3 over CE_1 and CE_2, since $W(a) = 7$. All these best solutions are obtained by using JaCoP.

6   Related Work

In [26], the authors have developed the notion of fuzzy unification and incorporated it into a novel fuzzy argumentation framework for extended logic programming: the attacks are associated to a fuzzy strength value, i.e. a V-attack. As well, a V-argument A is V-acceptable w.r.t. the set $Args$ of V-arguments if each argument V-attacked A is V-attacked by an argument in $Args$.

In [3], AFs have been also extended to *Value Based Argumentation Frameworks (VAF)* where V is a generic nonempty set of values and Val is a function which maps from elements of $Args$ to elements of V.

The work in [2] concerns the "acceptability" of arguments in preference-based argumentation frameworks. Preferences are represented with a preordering relationships (partial or total) that resembles the ordering defined by the $+$ operator of semirings (see Sec. 3).

```
// Defining the Variables of the SCSP
v[0] = new BooleanVariable(store, "a");
v[1] = new BooleanVariable(store, "b");
v[2] = new BooleanVariable(store, "c");
v[3] = new BooleanVariable(store, "d");
v[4] = new BooleanVariable(store, "e");

// conflict-free constraints
public static void imposeConstraintConflictFree(Store store, BooleanVariable[] v) {
store.impose(new ExtensionalConflictVA(new BooleanVariable[]{v[0], v[1]},
                                        new int[][]{{1, 1}}));
store.impose(new ExtensionalConflictVA(new BooleanVariable[]{v[2], v[1]},
                                        new int[][]{{1, 1}}));
store.impose(new ExtensionalConflictVA(new BooleanVariable[]{v[2], v[3]},
                                        new int[][]{{1, 1}}));
store.impose(new ExtensionalConflictVA(new BooleanVariable[]{v[3], v[2]},
                                        new int[][]{{1, 1}}));
store.impose(new ExtensionalConflictVA(new BooleanVariable[]{v[3], v[4]},
                                        new int[][]{{1, 1}}));
store.impose(new ExtensionalConflictVA(new BooleanVariable[]{v[4], v[4]},
                                        new int[][]{{1, 1}})); }

// stable constraints
public static void imposeConstraintStableExtensions(Store store, BooleanVariable[] v) {
store.impose(new ExtensionalConflictVA(new BooleanVariable[]{v[0]},
                                        new int[][]{{0}}));
store.impose(new ExtensionalConflictVA(new BooleanVariable[]{v[0], v[1]},
                                        new int[][]{{0, 0, 0}}));
store.impose(new ExtensionalConflictVA(new BooleanVariable[]{v[2], v[3]},
                                        new int[][]{{0, 0}}));

store.impose(new ExtensionalConflictVA(new BooleanVariable[]{v[3], v[4]},
                                        new int[][]{{0, 0}})); }
```

Fig. 6. The conflict-free and stable constraints in JaCoP for the mapping of Fig. 5

Probabilistic Argumentation [15,19]. This theory is an alternative approach for non-monotonic reasoning under uncertainty. It allows to judge open questions (hypotheses) about the unknown or future world in the light of the given knowledge. From a qualitative point of view, the problem is to derive arguments in favor and against the hypothesis of interest.

In [22] the author has extended Dung's theory of argumentation to integrate metalevel argumentation about preferences. Dung's level of abstraction is preserved, so that arguments expressing preferences are distinguished by being the source of a second attack relation that abstractly characterizes application of preferences by attacking attacks between the arguments that are the subject of the preference claims.

A close work is represented by [13]: there the authors introduce and investigate a natural extension of Dungs well known model of argument systems in which attacks are associated with a weight, indicating the relative strength of the attack. A key concept in that framework is the notion of an inconsistency budget, which characterizes how much inconsistency we are prepared to tolerate: given an inconsistency budget β, we would be prepared to disregard attacks up to a total cost of β.

Comparison. The framework proposed in this paper is able to solve all the above reported AFs (including the classical Dung framework [11]), both from the qualitative and (main novelty) quantitative point of view. Since in this paper we mainly propose a solving framework, we compare it with other related works.

In [13] weights are associated with attacks instead of arguments, as in our proposal. Moreover, no solving mechanism is proposed to solve the problems presented in the paper, even if their solution is proved to be difficult in the paper (e.g. *NP-Complete*). Moreover, in [13] the combination of the weights and the preference of the solution correspond to our *Weighted* semiring, while other possibilities are not considered.

In [18] crisp constraint have been used to model argumentation as constraint propagation in *Distributed Constraint Satisfaction Problem (DSCP)*. Different agents represent the distributed points in the problem. The paper shows the appropriateness of constraints in solving large-scale argumentation systems. However, it seems to only solve classical problems, (i.e. no qualitative or quantitative extensions).

The are some frameworks based on Logic Programming-like languages. For example, the system *ASPARTIX* [14] is a tool for computing acceptable extensions for a broad range of formalizations of Dung's argumentation framework and generalizations thereof, e.g. value-based AFs [3] or preference-based [2]. *ASPARTIX* relies on a fixed disjunctive datalog program which takes an instance of an argumentation framework as input, and uses the answer-set solver DLV for computing the type of extension specified by the user. However, *ASPARTIX* does not solve any quantitative argumentation case, as well as other Answer Set Programming systems [24].

In [9] the authors solve over-constrained weighted AF problems, where weights are associated with arcs and represent the cost of the attack between two arguments. to relax the notion of conflict-free extensions to α-conflict-free ones (and also for hte other extensions of Dung), in order to include in the same set also attacking arguments, whose attack costs are not worse than a threshold α.

7   Conclusions and Future Work

In the paper we have revised the notions provided by Dung [11] in order to associate the argument preference with a weight (taken from a semiring structure) that represents the "goodness" of the argument in terms of cost, fuzziness, probability or else. Further on, we have suggested the Dung's semantics in their soft version. Moreover, we have presented a mapping from SCSPs to AFs and solved the obtained SCSP with JaCoP, a Java Constraint Programming solver, thus finding the solution of the related AF. We have proposed an unifying computational framework with strong mathematical foundations and solving techniques, where by only parametrically changing the semiring we can deal with different weighted (or not) AFs. By having a uniform framework, it may be possible to see more clearly the relationships between different proposals. It may also offer the possibility to identify new results concerning classes of these proposals.

The user only needs to state the problem, while the underlying machinery is able to efficiently satisfy the constraints. Constraint solving techniques prove to be able to deal with large scale problems [18], even if the treated problems are difficult: for example, deciding if a set is a preferred extension is a *CO-NP*-complete problem [4]. Practical applications may consist, for example, in automatically study Discussion Fora where arguments are rated by users.

Notice that our soft constraint framework is able to solve all crisp and weighted extensions of Dung shown in Sec. 6 in a parametric way; therefore, the strength of this paper is to propose a general way to solve AFs.

In the future, we would like to cluster arguments according to their (for example) coherence, still using soft constraints as the framework to obtain the solution. This can be useful to check the discrepancies/likeness during a negotiation process, inside different interviews to the same political candidate or during discussions in general. As an example, *"We do not want immigrants with the right to vote"* is clearly closer to *"Immigration must be stopped"*, than to *"We need a multicultural and open society in order to enrich the life of everyone and boost our economy"*, and should belong to the same cluster.

At last, we want to generate a small-world network, for example with the *Java Universal Network/Graph Framework (JUNG)* [25] in order to test automatically give an interaction graph as input and test the related performance.

Acknowledgements

We would like to thank Massimiliano Giacomin for the important suggestions.

References

1. Amgoud, L., Cayrol, C., Lagasquie-Schiex, M.-C., Livet, P.: On bipolarity in argumentation frameworks. Int. J. Intell. Syst. 23(10), 1062–1093 (2008)
2. Amgoud, L., Cayrol, C.: Inferring from inconsistency in preference-based argumentation frameworks. J. Autom. Reasoning 29(2), 125–169 (2002)
3. Bench-Capon, T.J.M.: Persuasion in practical argument using value-based argumentation frameworks. J. Log. Comput. 13(3), 429–448 (2003)
4. Besnard, P., Doutre, S.: Checking the acceptability of a set of arguments. In: Workshop on Non-Monotonic Reasoning, pp. 59–64 (2004)
5. Bistarelli, S.: Semirings for Soft Constraint Solving and Programming. LNCS, vol. 2962. Springer, Heidelberg (2004)
6. Bistarelli, S., Montanari, U., Rossi, F.: Soft concurrent constraint programming. ACM Trans. Comput. Logic 7(3), 563–589 (2006)
7. Bistarelli, S., Montanari, U., Rossi, F.: Semiring-based Constraint Solving and Optimization. Journal of the ACM 44(2), 201–236 (1997)
8. Bistarelli, S., Santini, F.: Propagating multitrust within trust networks. In: ACM Symposium on Applied Computing, pp. 1990–1994. ACM, New York (2008)
9. Bistarelli, S., Santini, F.: A common computational framework for semiring-based argumentation systems. In: European Conference on Artificial Intelligence, ECAI 2010 (2010)

10. Coste-Marquis, S., Devred, C., Marquis, P.: Constrained argumentation frameworks. In: Knowledge Representation and Reasoning (KR), pp. 112–122. AAAI Press, Menlo Park (2006)
11. Dung, P.M.: On the acceptability of arguments and its fundamental role in non-monotonic reasoning, logic programming and n-person games. Artif. Intell. 77(2), 321–357 (1995)
12. Dunne, P.E., Hunter, A., McBurney, P., Parsons, S., Wooldridge, M.: Inconsistency tolerance in weighted argument systems. In: Conf. on Autonomous Agents and Multiagent Systems, pp. 851–858. IFAAMS (2009)
13. Dunne, P.E., Hunter, A., McBurney, P., Parsons, S., Wooldridge, M.: Inconsistency tolerance in weighted argument systems. In: Proceedings of The 8th International Conference on Autonomous Agents and Multiagent Systems, AAMAS 2009, pp. 851–858. International Foundation for Autonomous Agents and Multiagent Systems (2009)
14. Egly, U., Gaggl, S.A., Woltran, S.: ASPARTIX: Implementing argumentation frameworks using answer-set programming. In: Garcia de la Banda, M., Pontelli, E. (eds.) ICLP 2008. LNCS, vol. 5366, pp. 734–738. Springer, Heidelberg (2008)
15. Haenni, R.: Probabilistic argumentation. J. Applied Logic 7(2), 155–176 (2009)
16. Janssen, J., De Cock, M., Vermeir, D.: Fuzzy argumentation frameworks. In: Information Processing and Management of Uncertainty in Knowledge-based Systems, pp. 513–520 (2008)
17. Jøsang, A., Ismail, R., Boyd, C.: A survey of trust and reputation systems for online service provision. Decis. Support Syst. 43(2), 618–644 (2007)
18. Jung, H., Tambe, M., Kulkarni, S.: Argumentation as distributed constraint satisfaction: applications and results. In: Conference on Autonomous Agents (AGENTS), pp. 324–331. ACM, New York (2001)
19. Kohlas, J.: Probabilistic argumentation systems a new way to combine logic with probability. J. of Applied Logic 1(3-4), 225–253 (2003)
20. Kraus, S., Sycara, K., Evenchik, A.: Reaching agreements through argumentation: a logical model and implementation. Artif. Intell. 104(1-2), 1–69 (1998)
21. Kuchcinski, K., Szymanek, R.: Jacop - java constraint programming solver (2001), http://jacop.osolpro.com/
22. Modgil, S.: Reasoning about preferences in argumentation frameworks. Artif. Intell. 173(9-10), 901–934 (2009)
23. Montanari, U.: Networks of constraints: Fundamental properties and applications to picture processing. Inf. Sci. 7, 95–132 (1974)
24. Nieves, J.C., Cortés, U., Osorio, M.: Possibilistic-based argumentation: An answer set programming approach. In: Mexican International Conference on Computer Science (ENC), pp. 249–260. IEEE Computer Society, Los Alamitos (2008)
25. O'Madadhain, J., Fisher, D., White, S., Boey, Y.: The JUNG (Java Universal Network/Graph) framework. Technical report, UC Irvine (2003)
26. Schroeder, M., Schweimeier, R.: Fuzzy argumentation for negotiating agents. In: AAMAS, pp. 942–943. ACM, New York (2002)

Appendix A

The appendix shows all the JaCoP [21] code written to solve the AF_S proposed in Fig. 5.

```
package ExamplesJaCoP;

import JaCoP.constraints.ExtensionalConflictVA;
import JaCoP.core.*;
import JaCoP.search.*;
import java.util.ArrayList;
import java.util.Vector;

public class Argumentation {

    static Argumentation m = new Argumentation();
    static int size = 5;                        // number of variables
    static int[] weights = {7, 20, 6, 10, 12};     // weights associated with arguments
    static String[] labels = {"Conflict free", "Admissible sets", "Stable extensions",
            "Complete extensions", "Preferred Extensions", "Ground extensions"};
    static int set = 0;
    static Store store;             // store
    static BooleanVariable[] v;     // array of variables

    public static void main(String[] args) {
        // defining the store
        store = new Store();

        // defining the array of variables
        v = new BooleanVariable[size];

        // defining the single variable inside the store
        v[0] = new BooleanVariable(store, "a");
        v[1] = new BooleanVariable(store, "b");
        v[2] = new BooleanVariable(store, "c");
        v[3] = new BooleanVariable(store, "d");
        v[4] = new BooleanVariable(store, "e");

        /**
         * 0 = conflict free extensions
         * 1 = stable set extensions
         */
        switch (set) {

            case 0: // conflict free
                imposeConstraintConflictFree(store, v);
                break;

            case 1: // stable extensions
                imposeConstraintConflictFree(store, v);
                imposeConstraintStableExtensions(store, v);
                break;
        }

        /*
         * returning the solutions
         */
        getSolutions(store, v, set);
        System.out.println("");

    }

    // conflict-free constraints
    public static void imposeConstraintConflictFree(Store store, BooleanVariable[] v) {
        store.impose(new ExtensionalConflictVA(new BooleanVariable[]{v[0], v[1]},
                new int[][]{{1, 1}}));
        store.impose(new ExtensionalConflictVA(new BooleanVariable[]{v[2], v[1]},
```

```
                new int[][]{{1, 1}}));
    store.impose(new ExtensionalConflictVA(new BooleanVariable[]{v[2], v[3]},
                new int[][]{{1, 1}}));
    store.impose(new ExtensionalConflictVA(new BooleanVariable[]{v[3], v[2]},
                new int[][]{{1, 1}}));
    store.impose(new ExtensionalConflictVA(new BooleanVariable[]{v[3], v[4]},
                new int[][]{{1, 1}}));
    store.impose(new ExtensionalConflictVA(new BooleanVariable[]{v[4], v[4]},
                new int[][]{{1, 1}}));
}


// stable constraints
public static void imposeConstraintStableExtensions(Store store, BooleanVariable[] v) {
    store.impose(new ExtensionalConflictVA(new BooleanVariable[]{v[0]},
                new int[][]{{0}}));
    store.impose(new ExtensionalConflictVA(new BooleanVariable[]{v[0], v[2], v[1]},
                new int[][]{{0, 0, 0}}));
    store.impose(new ExtensionalConflictVA(new BooleanVariable[]{v[2], v[3]},
                new int[][]{{0, 0}}));
    // the constraint below is redundant w.r.t. the one just above
    //store.impose(new ExtensionalConflictVA(new BooleanVariable[]{v[3], v[2]},
                new int[][]{{0, 0}}));
    store.impose(new ExtensionalConflictVA(new BooleanVariable[]{v[3], v[4]},
                new int[][]{{0, 0}}));
}


public static void getSolutions(Store store, BooleanVariable[] v, int set) {

    // search for a solution and print results
    Search label = new DepthFirstSearch();
    // ordering the solutions
    SelectChoicePoint select = new InputOrderSelect(store, v, new IndomainMax());
    label.getSolutionListener().searchAll(true);
    // record solutions; if not set false
    label.getSolutionListener().recordSolutions(true);
    boolean result = label.labeling(store, select);
    int[][] solutions = label.getSolutionListener().getSolutions();


    if (set == 0 || set == 1) {
        // printing the solutions
        System.out.print(labels[set] + ": ");
        for (int i = 0; i < label.getSolutionListener().solutionsNo(); i++) {
            System.out.print("(");
            for (int j = 0; j < size; j++) {
                if (solutions[i][j] == 1) {
                    System.out.print(v[j].id);
                }
            }
            System.out.print(")");
        }

        // obtaining the best solutions
        Vector<Integer> bestSolutions = getBestSolutions(solutions,
                label.getSolutionListener().solutionsNo());
        // printing the best solutions
        printBestSolutions(bestSolutions, solutions);


    }
}


// array as in input and returns the indexes of the best elements
// computing solutions with the best (i.e. lowest) cost
public static Vector<Integer> getBestSolutions(int[][] solutions, int solutionNumber) {
```

```
        Vector<Integer> solutionIndexes = new Vector<Integer>();
        Integer[] solutionWithWeight = new Integer[2];

        int min = Integer.MAX_VALUE;
        int solutionWeight = 0;

        for (int j = 0; j < solutionNumber; j++) {
            solutionWeight = 0;
            solutionWithWeight[0] = 0;
            solutionWithWeight[1] = 0;

            for (int c = 0; c < size; c++) {
                if (solutions[j][c] == 1) {
                    solutionWeight = solutionWeight + weights[j];
                }
            }

            if (solutionWeight < min) {
                solutionIndexes.removeAllElements();
                min = solutionWeight;
                solutionIndexes.add(j);
                //System.out.println("index of the added solution " + j);
            } else if (solutionWeight == min) {
                solutionIndexes.add(j);
            }

        }

        //System.out.println("solution index: " + solutionIndexes.get(0));
        return solutionIndexes;
    }

    public static int getWeigthSolution(int[] solution) {
        int weigthSolution = 0;
        for (int i = 0; i < size; i++) {
            if (solution[i] == 1) {
                weigthSolution = weigthSolution + weights[i];
            }
        }
        return solutionWeight;
    }

    public static void printBestSolutions(Vector<Integer> bestSolutions, int[][] solutions) {
        System.out.println("");
        System.out.print("Bests " + labels[set] + ": ");
        for (int i = 0; i < bestSolutions.size(); i++) {
            System.out.print("(");
            for (int j = 0; j < size; j++) {
                if (solutions[bestSolutions.get(i)][j] == 1) {
                    System.out.print(v[j].id);
                }
            }
            System.out.print(") = " + getWeigthSolution(solutions[bestSolutions.get(i)]));
        }
    }
}
```

Connecting BnB-ADOPT with Soft Arc Consistency: Initial Results*

Patricia Gutierrez and Pedro Meseguer

IIIA, Institut d'Investigació en Intel.ligència Artificial
CSIC, Consejo Superior de Investigaciones Científicas
Campus UAB, 08193 Bellaterra, Spain
{patricia,pedro}@iiia.csic.es

Abstract. Distributed constraint optimization problems with finite domains can be solved by asynchronous procedures. ADOPT is the reference algorithm for this kind of problems. Several versions of this algorithm have been proposed, one of them is BnB-ADOPT which changes the nature of the original algorithm from best-first to depth-first search. With BnB-ADOPT, we can assure in some cases that the value of a variable will not be used in the optimal solution. Then, this value can be deleted unconditionally. The contribution of this work consists in propagating these unconditionally deleted values using soft arc consistency techniques, in such a way that they can be known by other variables that share cost functions. When we propagate these unconditional deletions we may generate some new deletions that will also be propagated. The global effect is that we search in a smaller space, causing performance improvements. The effect of the propagation is evaluated on several benchmarks.

1 Introduction

A classical problem in computer science is finding a global optimum of an aggregation of some elementary cost functions. Many real life problems can be represented as a collection of constraints or penalty relations over a set of variables. A constraint optimization algorithm is a solver able to find an assignment to every variable that satisfies all the constraints or, in case not all constraints can be satisfied, it reduces the total cost finding the minimum penalty for the resulting variable assignment. When the variables and constraints of the problem are not centralized and the information is distributed among several automated agents the problem is a *Distributed Constraint Optimization Problem* (DCOP). Cost functions can be distributed for several reasons: privacy issues, distributed origin of the problem data, high translation costs into a centralized setting, etc. Distributed resolution involves message passing among agents holding the cost functions and the involved variables. Agents must cooperate to find the global optimum (minimum) cost. DCOPs can be found in many real domains for modeling a variety of multiagent coordination problems such as distributed planning, scheduling, sensor networks and others.

In distributed search, the first complete algorithm for DCOPs was ADOPT [6], which has evolved producing different versions as BnB-ADOPT [8] and ADOPT-ng [7].

---

* This work is partially supported by the project TIN2009-13591-C02-02.

J. Larrosa and B. O'Sullivan (Eds.): CSCLP 2009, LNAI 6384, pp. 19–37, 2011.

BnB-ADOPT changes the search strategy of the original ADOPT from best-first to depth-first. BnB-ADOPT offers a better performance than ADOPT, it keeps ADOPT good theoretical properties and requires a relatively simple implementation.

In BnB-ADOPT there are cases when a value of a variable will not be used in the optimal solution. Then, this value can be deleted unconditionally. The contribution of this work is to remove those values from the DCOP instance, and propagate these unconditionally deleted values in such a way that they can be known by other variables. This is useful because when we propagate unconditional deletions we may generate some new deletions that will also be propagated (which may generate further deletions, etc). Propagation is done using soft arc consistency methods, which are adequate for this kind of problems. This novel combination generates the BnB-ADOPT-AC* algorithm. As the domain of the variables is being reduced and the search space becomes smaller, the performance of BnB-ADOPT-AC* improves over BnB-ADOPT, in terms of the number of messages exchanged or the number of cycles required to achieve the optimum. Experimental results clearly indicate that this approach pays-off, causing significant reductions in the communication cost, and in the number of cycles required.

This paper is organized as follows. In section 2, we present existing approaches for the centralized and distributed cases of constrained optimization. For the centralized case, we present the COP definition and some basics of soft arc consistency techniques specific for the weighted model. For the distributed case, we present the DCOP definition with a description of BnB-ADOPT. In section 3 we connect soft arc consistency with BnB-ADOPT, showing how these consistencies can be integrated in the BnB-ADOPT messages (one extra message is required). Using them, we propagate unconditionally deleted values. An example of the new algorithm including its trace appears in section 4. The practical benefits of connecting BnB-ADOPT with soft arc consistency become apparent in section 5, where the original and the new algorithm are compared on two diferent benchmarks. The new algorithm substantially decreases the number of exchanged messages and the number of cycles required to solve tested DCOP instances. Finally, section 6 contains some conclusions of this work and lines for further research.

2 Preliminaries

2.1 Centralized Case

Distributed Constraint Optimization Problem. A *Constraint Optimization Problem (COP)* involves a finite set of variables, each taking a value in a finite domain [1]. Variables are related by cost functions that specify the cost of value tuples on some variables subsets. Costs are positive natural numbers (including 0 and ∞). A finite *COP* is a tuple (X, D, C) where:

- $X = \{x_1, \ldots, x_n\}$ is a set of n variables;
- $D = \{D(x_1), \ldots, D(x_n)\}$ is a collection of finite domains; $D(x_i)$ is the initial domain of x_i;
- C is a set of cost functions; $f_i \in C$ specifies the cost of every combination of values of $var(f_i)$ on the ordered set of variables $var(f_i) = (x_{i_1}, \ldots, x_{i_{r(i)}})$, that is:

$f_i : \prod_{j=i_1}^{i_{r_i}} D(x_j) \mapsto N \cup \{0, \infty\}.$
The arity of f_i is $|var(f_i)|$.

The overall cost of a complete tuple (involving all variables) is the addition of all individual cost functions evaluated on that particular tuple. A *solution* is a complete tuple with acceptable overall cost, and it is *optimal* if its overall cost is minimal. Clearly, this is an instance of the weighted model for soft constraints [5].

Soft Arc Consistency. Let $P = (X,D,C)$ be a binary *COP*, (i,a) the notation for variable i and value a, $\top$ the lowest unacceptable cost and $\bot$ the minimum allowed cost for the problem. As [2], we consider the following local consistencies for the weighted case:

- **Node consistency (NC)**: (i,a) is node consistent (NC) if $C_i(a) < \top$. Variable i is NC if all of its values are NC. A COP is NC if every variable is NC.
- **Arc consistency (AC)**: (i,a) is arc consistent (AC) with respect to constraint C_{ij} if it is NC and there is a value $b \in D_j$ such that $C_{ij}(a,b) = \bot$. Value b is called a *support* of a. Variable i is AC if all its values are AC with respect to every binary constraint affecting i. A COP is AC if every variable is AC.

Notice that when (i,a) is not NC we can remove a from the problem, since $C_i(a) \geq \top$ we can assure that any assignment containing value a for variable i will cost at least $\top$, so it will not be an acceptable solution.

AC in the weighted case can be enforced applying two basic operations until AC condition is satisfied: forcing supports to node-consistent values, and pruning node inconsistent values. Support can be forced by sending (projecting) the minimum cost from the binary constraints of a value to its unary constraint. The projection of the binary constraint C_{ij} over the unary constraint C_i for the value a is a flow of costs defined as follows: Let α_a be the minimum cost of a with respect to C_{ij} (namely $\alpha_a = min_{b \in D_j} C_{ij}(a,b)$). The projection consists in adding α_a to $C_i(a)$ (namely, $C_i(a) = C_i(a) + \alpha_a, \forall a \in D_i$) and subtracting α_a from $C_{ij}(a,b)$ (namely, $C_{ij}(a,b) = C_{ij}(a,b) - \alpha_a, \forall b \in D_j, \forall a \in D_i$). This operation can be seen as if we were assuring that, no matter which value variable j will take, i will always have to pay α_a for value a.

It is worth noting that the systematic application of these two operations does not change the optimum cost and maintains an optimal solution. On one hand, constraint projection when applied to a problem $P = (X, D, C)$, produces an equivalent problem $P' = (X', D', C')$ ($X = X', D = D'$ and the same complete value tuple has the same cost in P and P'). On the other hand, although value pruning does not conserve problem equivalence, it produces a new problem with the same optimum as the original one. These properties are proved in [3].

Notice that when we prune a value from variable i to ensure AC, we need to recheck AC over every variable that i is constrained with, since the deleted value could be the *support* of the neighbor variable. Therefore, a deleted value in one variable might cause further deletions in other variables as a result of AC enforcement. The AC check must be performed until no further values are deleted.

We consider a stronger definition for node consistency and arc consistency [2]. In general, we can see the minimum cost of all unary constraints in variable x_i as the cost this variable will necessarily have to pay no matter which will be its assignment in the final solution. In this same way, all variables from the problem could project there minimum unitary cost over a zero-ary cost function C_ϕ, producing a necessary global cost of any complete assignment. So, we can point to an alternative definition of node consistency noted NC*, where we have a constant C_ϕ initially set to $\bot$. The idea is to project unary constraints over C_ϕ, which will become a global lower bound of the problem solution.

- **Node Consistency* (NC*):** (i,a) is node consistent* (NC*) if $C_\phi + C_i(a) < \top$. Variable i is NC* if: all its values are NC* and there exists value $a \in D_i$ such that $C_i(a) = \bot$. Value a is a *support* for the variable NC*. A COP is NC* if every variable is NC*.
- **Arc consistency* (AC*):** (i,a) is arc consistency* (AC*) with respect to constraint C_{ij} if it is NC* and there is a value $b \in D_j$ such that $C_{ij}(a,b) = \bot$. Value b is called a *support* of a. Variable i is AC* if all its values are AC* with respect to every binary constraint affecting i. A COP is AC* if every variable is AC*.

This AC* definition simply replaces NC by NC* in the previous AC definition. Enforcing AC* is a slightly more difficult task than enforcing AC, because C_ϕ has to be updated after the projection of binary constraints over unary constraints, and each time is updated all domains must be revised for new node-inconsistent values.

In the following, we restrict our work to the weighted model of soft constraints.

2.2   Distributed Case

Distributed COP. A *Distributed Constraint Optimization Problem(DCOP)*, is a *COP* where variables, domains and cost functions are distributed among automated agents. Formally, a *DCOP* is a 5-tuple (X, D, C, A, α), where X, D, C define a *COP* and:

- $A = \{1, \ldots, p\}$ is a set of p agents
- $\alpha : X \rightarrow A$ maps each variable to one agent

Clearly, DCOP significantly generalizes the Distributed Constraint Satisfaction Problem (*DisCSP*) framework in which problem solutions are characterized with a designation of satisfactory or unsatisfactory (in the final solution all constraints must be satisfied) and do not model problems where solutions have degrees of quality cost.

For simplicity, we assume that each agent holds exactly one variable (so the terms variable and agents can be used interchangeably) and cost functions are unary and binary only (in the following a cost function will be denoted as C with the indexes of variables involved, so C_{ij} is the binary constraint between agent i and j, and C_i is a unary constraint over agent i). A constraint is known and can be accessed by any agent that is constrained by it, but not for other agents. To coordinate agents will need to exchange messages: it is assumed that these messages can have a finite delay but they will eventually be received, and for a given pair of agents messages are delivered in the order they were sent.

Although soft arc consistency concepts were introduced in a centralized setting, they are also applicable to a distributed context without any difference. In the following, we use these concepts inside algorithms for DCOP solving.

BnB-ADOPT (Branch and Bound ADOPT). ADOPT [6] was the first complete algorithm for asynchronous distributed constraint optimization. ADOPT allows each agent to change its value whenever it detects a better local assignment. With this strategy, it can find the DCOP optimum, or a solution within a user-specified distance from the optimum, with local communication and polynomial space at each agent.

Aiming at improving ADOPT performance, several ADOPT-based algorithms have been proposed, ADOPT-ng [7] and BnB-ADOPT [8]. On BnB-ADOPT, while ADOPT explores the search tree in best-first order, BnB-ADOPT explores the search tree in depth-first order, testing the children of a node in increasing cost order and pruning those nodes whose cost is greater than a given *threshold*. This strategy makes it memory bounded without having to repeatedly reconstruct partial solutions previously discarded. BnB-ADOPT assumes that an agent can neither observe the cost of constraints that it is not involved in nor the values that other agents take on, and agents can exchange messages with their neighbors only.

Like ADOPT, agents in BnB-ADOPT are arranged in a *DFS* tree [8]. It also uses the message passing and communication framework of ADOPT, using three message types: VALUE, COST and STOP. A generic agent *self* sends a VALUE message when it changes value, to inform children and pseudo-children. In response, children send COST messages to *self* informing the cost of this new assignment calculated on their instances with the current information. Finally, STOP messages are sent when the optimum is found, or when an agent discovers that there is no solution. Agent *self* maintains the following data structures:

- The current context: the set of (variable,value) assignments representing what *self* believes of the assignments of higher agents in its branch of the *DFS* tree. At the beginning *self* has an empty context, during search this context is updated when an ancestor sends to *self* a VALUE message. Two context are *compatible* if they do not disagree in any assignment of common variables.
- For each value $v \in D(self)$ and each child *child*, it holds: a lower bound $lb(v, child)$, an upper bound $ub(v, child)$, and a $context(v, child)$. These tables are updated with the information sent by children in COST messages. If at some point the context of *self* becomes incompatible with the context of *child*, the information sent by *child* is considered obsolete and $lb(\_, child)$, $ub(\_, child)$ and $context(\_, child)$ tables are reinitialized.

With them *self* calculates its own lower bound *LB* and upper bound *UB*, based on its local cost plus any cost reported by its children:

$$LB[v] \leftarrow \sum_{(x_i, v_i) \in myContext} C_{self, x_i}(v, v_i) + \sum_{x_k \in myChildren} lb[v, x_k]$$

$$UB[v] \leftarrow \sum_{(x_i, v_i) \in myContext} C_{self, x_i}(v, v_i) + \sum_{x_k \in myChildren} ub[v, x_k]$$

$$LB \leftarrow min_{v \in D_{self}} LB[v]; \qquad UB \leftarrow min_{v \in D_{self}} UB[v]$$

Finally, *self* also maintains a threshold *TH* initialized to ∞ (and in the *root* agent remains always ∞) used for pruning during depth-first search. This value is iteratively refined as new global upper bounds are found.

At the beginning of the execution, every agent chooses the value that minimizes its *LB* and sends a VALUE message to its children and pseudo-children informing the new assignment. When *self* receives a VALUE message, it updates its context with the new assignment and checks if the updated context remains compatible with the children contexts: if they are not compatible the information provided by that child (stored in *lb*, *ub* and *context* tables) is treated as obsolete and tables are reinitialized. In VALUE messages the *TH* value is also propagated (initially ∞). When *self* receives a COST message, the sending child informs of its *LB* and *UB*. As these bounds are calculated depending on values of higher variables, the child must attach the context under which these costs are calculated. If the sent context is compatible with its own context, *self* updates the information received in the *lb*, *ub* and *context* tables. Also, *self* includes into its own context any higher variable assignment that does not share a constraint with *self*. This is because any change on these variables may cause that the stored *lb* and *ub* for this child become obsolete, so *self* stores the higher variable to detect that.

Once every message is processed, *self* decides if it must change its value. If the *LB* of the current value is greater than or equal to *min(TH, UB)*, *self* will change its value to the one with smallest *LB*. This means that, for the current context, each agent will maintain its value until its cost becomes greater or equal that the local upper bound. In that case the value is proven to be suboptimal and it can be discarded. The condition holds during the current context only, since the *UB* is reinitialized to ∞ every time the context changes involving a child context. Notice that on first iterations the *UB* for an agent will be infinite until every child has informed with its corresponding COST (so an agent will not change its value until all information from its children is received), but on later iterations it may not be necessary to wait for every child to inform its cost, since an agent can decide to discard the current value only with partial information if the *LB* of its value is already greater than the stored *TH*, which works as a global upper bound for that child subtree.

After the agent has decide if it must change its value, it sends a VALUE message to each child *child* with its current assignment and the desired threshold, calculated as:

$$min(TH,UB) - \sum_{\substack{(x_i, d_i) \in \\ myContext}} C_{self,x_i}(myValue, d_i) - \sum_{\substack{x_j \in myChildren \\ x_j \neq child}} lb[myValue, x_j]$$

This desired threshold is used for pruning when a child reaches this value. It can be seen as an estimated upper bound for the child subtree during the current context, since we are removing from the parent upper bound its local cost and the informed *lb* of all children except for the one the message is been sent to. In the next step the termination condition is checked, this condition is triggered by the *root* when *LB = UB*. This condition can only be achieved when all values in the *root* domain have been explored or pruned. Finally, a COST message is sent to the parent informing of the *LB* and *UB* costs under the current context.

3   Connecting BnB-ADOPT with Soft Arc Consistency

In this section we present our contribution combining the search strategy of BnB-ADOPT with the inference technique enforcing AC* over the DCOP instance. Due to the distributed setting this combination requires some care. In a naive approach, each time an agent needs information of other agents it would generate two messages (the request and response) which could cause a serious degradation in performance. In our approach, we try to keep the number of exchanged messages as low as possible, introducing the required elements to enforce AC* in the existing BnB-ADOPT messages, keeping their meanings for distributed search. Only a new type of message is added.

3.1   Propagating Unconditional Deletions

Let us consider a DCOP instance, where agents are arranged in a DFS tree and each executes BnB-ADOPT. Imagine the $root$ agent, $D_{root} = \{a, b, \ldots\}$. Let us assume that it takes value a. After a while, $root$ will know $cost(a) = lb(a) = ub(a)$, and it decides to change its assignment to b, informing to its children with the corresponding VALUE messages. Children start answering about the cost of b (this is a change of context in BnB-ADOPT terms) with COST messages. As soon as $root$ realizes that $cost(b) > cost(a)$, b can be removed from D_{root} since it will not be in the solution and it will never be considered again (a similar situation happens if $cost(a) > cost(b)$, then a can be removed from D_{root}). Just removing b from D_{root} will cause no effect in BnB-ADOPT, because it will not consider b again as possible value for $root$. However, if we inform constrained agents that b is no longer in D_{root}, this may cause some values of other agents to become unfeasible so they can be deleted as well.

A related situation happens when an internal node of the DFS-tree, agent $self$, receives COST messages from its children. A COST message contains the lower bound computed by BnB-ADOPT, with the context (variable, value) pairs on which this lower bound was computed. Let us consider COST messages whose context is simply the $self$ agent with its actual value v. If the sum of the lower bounds of these COST messages reaches or exceeds $\top$, the value v of $self$ can be deleted. To see this, it is enough to realize that the lower bound is computed assuming (variable, value) pairs of context: if this is simply $(self, v)$, the actual cost of v does not depend on the value of any other agent, so if it reaches or exceeds $\top$ it can be deleted.

In these two cases, deletions are unconditional: (i) $root$ values does not depend on any context (there is no higher agent), and (ii) at an internal node $self$, we consider the case of contexts that depend on $self$ only. These deletions can be further propagated in the same way, decrementing the size of the search space. Any deletion caused by propagation of unconditional deletions is also unconditional.

To propagate these value deletions to other agents we need to maintain soft arc consistency in the distributed instance. Let us consider variables x_i and x_j with unitary costs C_i and C_j, and constrained by cost function C_{ij}. We consider the NC* and AC* notions defined in section 2.1 following [2]. To apply this idea to DCOPs, we assume that the distributed instance is initially AC* (otherwise, it can be made AC* by preprocessing). If value a is unconditionally deleted from D_i, it might happens that value $a \in D_i$ were the only support of a value $b \in D_j$. For this reason, after the notification of

BnB-ADOPT messages:
VALUE($sender, destination, value, threshold$)
COST($sender, destination, context[], lb, ub$)
STOP($sender, destination$)

BnB-ADOPT-AC* messages:
VALUE($sender, destination, value, threshold, \top, C_\phi$)
COST($sender, destination, context[], lb, ub, subtreeContribution$)
STOP($sender, destination, emptydomain$)
DEL($sender, destination, value$)

Fig. 1. Messages of BnB-ADOPT and BnB-ADOPT-AC*

a deletion, directional AC* has to be enforced on cost function C_{ij} from i to j (observe that in the other direction enforcing will cause no change). This has to be done in both agents i and j, to assure that both maintain the same representation of C_{ij}. In addition, agent j may pass binary costs to unary costs, which might result that some value $b \in D_j$ becomes not NC*. In that case, b should be deleted from D_j and its deletion should be propagated in the same way.

3.2   BnB-ADOPT-AC*

The idea of propagating unconditional deletions can be included in BnB-ADOPT, producing the new BnB-ADOPT-AC*, where the semantic of original BnB-ADOPT messages remains unchanged. New elements are included in these messages: changes with respect to the original ones appear in Figure 1. BnB-ADOPT-AC* requires some minor changes with respect to BnB-ADOPT:

– In addition to its own domain, the domain of every variable constrained with *self* is also represented in *self*. The binary constraints between any pairs of agents will be represented in both agents, they are assumed to be AC*.
– A new message type, DEL, is required. When $self$ deletes value a in $D(self)$, it sends a DEL message to every agent constrained with it. When $self$ receives a DEL message, it registers that the message value has been deleted from the domain of sender, and it enforces AC* on the constraint between $self$ and sender. If, as result of this enforcing, some value is deleted in $D(self)$ it is propagated as above.
– VALUE messages include $\top$ and C_ϕ. $\top$ is initially ∞ and when $root$ reaches its first global upper bound $\top$ becomes a finite value which is propagated downwards, informing the other agents of the lowest unacceptable cost. Contributions to C_ϕ are propagated upwards in COST messages and aggregated in $root$, forming C_ϕ, the minimum global cost of the instance (no matter the values assigned to variables). Then, C_ϕ is propagated downwards in VALUE messages.
– COST messages include the contribution of each agent to the global C_ϕ. Each agent adds its own contribution with the contributions of all its children, and the result is included in the next COST message sent to its parent. All these contributions are finally added in $root$, forming the global C_ϕ, which is propagated donwards in VALUE messages. When there is a deletion C_ϕ might change, this is propagated.
– A value $a \in D(x_i)$ that satisfies the deletion condition is immediately deleted.

Maintaining AC* exploits the idea that a constraint C_{ij} is known by both agents i and j. Both agents keep a copy of the domain of others agent variable and the most updated representation of C_{ij}, with the purpose of generating as many deletions as possible. If value a is deleted in $D(x_i)$ (which is done by agent i) and the resulting domain is not empty, this deletion is notified to agent j. Then, agents i and j perform the same process, directional AC* from j to i, which assures that both maintain the same representation of C_{ij}. If in this process some value of $D(x_j)$ should be deleted, this is done by agent j, which notifies i and the same process repeats.

Theorem 1. *BnB-ADOPT-AC* computes the optimum cost and terminates.*

Proof. BnB-ADOPT performs a distributed depth-first traversal of the search tree defined by the DCOP instance to be solved. BnB-ADOPT-AC* does the same, with the only difference that, in some cases, variable domains may be smaller than initial domains. Messages trigger the same actions as in BnB-ADOPT, they simply include some extra information and check values for deletion. While BnB-ADOPT performs distributed search, in addition BnB-ADOPT-AC* also enforces AC*. This may cause new DEL messages which may only cause further DEL messages.

A value v is removed from D_{self} because its lower bound cost surpasses $\top$ (when *self = root* or *self* is an internal node). The lower bound cost of v is computed including lower bounds of children whose contexts mention *self* only (the lower bound calculated does not depend on any other high variable assignment). It is direct to check that v will not be in the optimal solution (otherwise, that solution will cost more than $\top$), so v can be removed from D_{self}. Enforcing AC*, values are removed using the projection and pruning operators of the weighted case of soft arc consistency. It is proven [3] that these operators do not change the optimum cost of the original instance.

If we would have known those values which are unconditionally not AC* prior starting BnB-ADOPT execution, we could remove them from their corresponding domains and speed-up BnB-ADOPT performance. However, since $\top$ and C_ϕ are evolving during distributed search, removing those values is not possible at the beginning. We will see that removing those values during algorithm execution has no effect in optimality and termination. Let us consider $v \in D_{self}$ that is going to be removed. If v is not the current assignement of x_{self}, *self* simply will remove v from D_{self} and inform its children and pseudochildren of it (via DEL messages). If v is the current assignement of x_{self}), removing v will not cause any extra difficulty in algorithm execution: since BnB-ADOPT-AC* is an asynchronous algorithm, any agent can change its value at any time. After removing v, BnB-ADOPT-AC* will follow the same search strategy as BnB-ADOPT, selecting as new value the one that minimizes the lower bound of available values in D_{self}. The termination condition remains unaltered. Since BnB-ADOPT computes the optimum cost and terminates [8] BnB-ADOPT-AC* also does so.   □

3.3   Preprocess Code

It is mentioned above that the initial problem is assumed to be AC*. If not, this can be easily done by preprocess executed on each agent, depicted in Figure 2.

procedure AC^*-preprocess($\top$)
  initialize;
  $AC^*(\top)$;
  while ($\neg end$) **do**
    $msg \leftarrow$ getMsg();
    switch($msg.type$)
      DEL: ProcessDelete(msg); $STOP$: ProcessStop(msg);

procedure $AC^*(\top)$
  for each $j \in neighbors(self)$ **do**
    if $j > self$ **then**
      FromBinaryToUnary($self, j$);
      FromUnaryToC$_\phi$($self$);
      FromBinaryToUnary($j, self$);
    else
      FromBinaryToUnary($j, self$);
      FromBinaryToUnary($self, j$);
      FromUnaryToC$_\phi$($self$);
  PruneDomainSelf($\top$);

procedure FromBinaryToUnary(i, j)
  for each $a \in D_i$ **do**
    $v \leftarrow argmin_{b \in D_j}\{C_{ij}(a,b)\}$; $\alpha \leftarrow C_{ij}(a,v)$;
    for each $b \in D_j$ **do** $C_{ij}(a,b) \leftarrow C_{ij}(a,b) - \alpha$;
    if $i = self$ **then** $C_i(a) \leftarrow C_i(a) + \alpha$;

procedure FromUnaryToC$_\phi$(i)
  $v \leftarrow argmin_{a \in D_i}\{C_i(a)\}$; $\alpha \leftarrow C_i(v)$;
  $myContribution \leftarrow myContribution + \alpha$;
  for each $a \in D_i$ **do** $C_i(a) \leftarrow C_i(a) - \alpha$;

procedure PruneDomainSelf($\top$)
  for each $a \in D_{self}$ **do if** $C_{self}(a) + C_\phi \geq \top$ **then** DeleteValue(a);

procedure DeleteValue(a)
  $D_{self} \leftarrow D_{self} - \{a\}$;
  FromUnaryToC$_\phi$($self$);
  if $D_{self} = \emptyset$ **then**
    for each $j \in neighbors(self)$ **do** sendMsg:$(STOP, self, j, true)$;
    $end \leftarrow true$;
  else for each $j \in neighbors(self)$ **do**
    sendMsg:$(DEL, self, j, a)$;
    FromBinaryToUnary($j, self$);

procedure ProcessDelete(msg)
  $D_{sender} \leftarrow D_{sender} - \{msg.value\}$;
  FromBinaryToUnary($self, sender$);
  FromUnaryToC$_\phi$($self$);
  PruneDomainSelf($\top$);

procedure ProcessStop(msg)
  if ($msg.emptyDomain = true$) **then**
    for each $j \in neighbors(self), j \neq sender$ **do** sendMsg($STOP, self, j, true$);
  $end \leftarrow true$;

Fig. 2. The preprocess algorithm for enforcing AC^*

After preprocess we will ensure the AC^* condition. Initially $C_\phi = 0$. Contributions to C_ϕ travel in COST messages, C_ϕ itself travels in VALUE messages, but these messages are not used in the preprocess (they implement distributed search), so C_ϕ remains 0 during the preprocess. So the level of soft arc consistency reached is actually AC.

This does not cause any harm, since after preprocess binary cost functions are projected on unary ones. The NC* condition (with $C_\phi > 0$) will be tested during distributed search.

- AC*-preprocess. It performs AC* on *self*. This may cause deletions so DEL messages are processed until no further deletions are generated.
- AC*. It performs the projection from binary to unary costs, executing procedure FromBinaryToUnary on all neighbors of *self*, plus the projection from unary to zero-ary costs in *self* executing FromUnaryToC$_\phi$. Projecting binary into unary costs is done orderly, projecting always first over the higher agent and then over the lower one in the *DFS* tree, to maintain a coherent copy of the cost functions on every agent. PruneDomainSelf(⊤) checks the NC* property on D_{self}.
- FromBinaryToUnary(i,j). It projects binary costs C_{ij} on unary costs C_i. When the projection is done over *self* the unary costs are stored in the agent.
- ProjectOverC$_\phi$(i). It projects unary costs on zero-ary costs in *myContribution*, which accumulates the contribution of *self* to the global C_ϕ.
- PruneDomainSelf. It checks every value in D_{self} for deletion, enforcing the NC* property.
- DeleteValue(a). *self* removes value a from D_{self}. If $D_{self} = \emptyset$, there is no acceptable solution, so STOP messages are sent to all neighbors, indicating that there is an empty domain and the process terminates. Otherwise, for all neighbors j, a DEL message is sent notifying a deletion and directional AC* is enforced; this operation is also done on every neighbor when the DEL message arrives.
- ProcessDelete(j,a). *self* has received a DEL message, notifying that agent j has deleted value a from D_j, so *self* registers this in its D_j copy. Then, it enforces directional AC* in $C_{self,j}$. Finally, *self* checks if it can prune some value from its domain executing PruneDomainSelf(⊤).
- ProcessStop(). *self* has received a STOP message. If it has been caused by an empty domain, *self* resends the STOP message to all its neighbors, except *sender*. Otherwise, it records the reception of the STOP message.

3.4   BnB-ADOPT-AC* Code

In order to use the stronger AC* condition in the process phase we need to maintain up to date the global C_ϕ. So every time there is a deletion and a call to the method FromBinaryToUnary, the minimum value from C_i is projected and propagated. In this way, the C_ϕ summarizes the obligatory costs from all the variables, not just from the one currently under investigation. Consequently, when we check the AC* condition there are more possibilities for pruning.

The process of BnB-ADOPT-AC* appears in Figures 3 and 4. The rationale for the changes is as follows. We assume that if $argmin_{v \in D_{self}}$ LB(v) is called with $D_{self} = \emptyset$, nil is returned. The main changes appears in ProcessValue when the C_ϕ is updated and a better ⊤ is founded (3 last lines), in ProcessCost when the subtree contribution to C_ϕ is calculated (5 last lines), and in the Backtrack procedure when a value is found suboptimal and deleted.

procedure BnB-ADOPT-AC$\star$()
  initialize tables and bounds;
  InitSelf(); Backtrack();
  while ($\neg end$) **do**
    while message queue is not empty **do**
      $msg \leftarrow$ getMsg();
      switch($msg.type$)
        $VALUE$:ProcessValue(msg); $COST$:ProcessCost(msg);
        $STOP$: ProcessStop; DEL: ProcessDelete(msg);
    Backtrack();

procedure ProcessValue(msg)
  if ($myContext[sender] \neq msg.value$) **then**
    $myContext[sender] \leftarrow msg.value$;
    CheckCurrentContextWithChildren(); InitSelf();
  if $sender = myParent$ **then** $TH \leftarrow msg.threshold$;
  if $msg.\top < \top$ **then** $\top \leftarrow msg.\top$;
  if $C_\phi < msg.C_\phi$ **then** $C_\phi \leftarrow msg.C_\phi$;
  PruneDomainSelf($\top$)

procedure ProcessCost(msg)
  $contextChange \leftarrow$ false;
  for each $x_i \in msg.context, x_i \notin myNeighhbors$ **do**
    if $myContext[x_i] \neq msg.context[x_i]$ **then**
      $myContext[x_i] \leftarrow msg.context[x_i]$; $contextChange \leftarrow$ true;
  if $contextChange =$ true **then** CheckCurrentContextWithChildren();
  if isCompatible($myContext, msg.context$) **then**
    $lb[msg.context[self], sender] \leftarrow msg.lb$;
    $ub[msg.context[self], sender] \leftarrow msg.ub$;
    $tableContext[msg.context[self], sender] \leftarrow msg.context$;
  if $contextChange =$ true **then** InitSelf();
  $childrenContribution[sender] \leftarrow msg.subtreeContribution$;
  $mySubtreeContr = myContribution$;
  for each $x_i \in myChildren$ **do**
    $mySubtreeContr = mySubtreeContr + childrenContribution[x_i]$;
  if $mySubtreeContr > C_\phi$ **then** $C_\phi \leftarrow mySubtreeContr$;

procedure InitSelf()
  $myValue \leftarrow argmin_{v \in D_{self}}$ LB(v); $TH \leftarrow \infty$;

Fig. 3. BnB-ADOPT-AC*, missing procs appear in Fig 2

- BnB-ADOPT-AC$\star$. Includes the reception of DEL messages: this message indicates that a value has been deleted in the domain of the sender. When received, the ProcessDelete procedure is called.
- InitSelf. The agent takes the value that minimize the LB and initializes its *threshold* to ∞.
- ProcessValue. The VALUE message includes the current $\top$ and the global C_ϕ, which are updated. *self* checks if it can prune some values from its domain with the received $\top$. When this message comes from $parent(self)$ the threshold is copied.
- ProcessCost. The COST message includes the children contribution to the global C_ϕ. *self* records this information and calculates the variable *mySubtreeContr* accumulating its own contribution plus the contribution of its children. If this is greater than the current C_ϕ then the global C_ϕ is updated.
- Backtrack. Depending wheter *self* is the *root* or not, it calls the procedure CheckForRootDeletions or CheckForInternalDeletions, which may also trigger unconditional deletions. Then, lower neighbors are informed of *self* value, and the termination condition is tested. Finally, when the COST

procedure Backtrack()
 if $TH \leq$ LB() **then** $TH \leftarrow \infty$;
 if LB($myValue$) $\geq min\{TH,$ UB()$\}$ **then**
   if $self = root$ **then** CheckForRootDeletions();
   else CheckForInternalDeletions();
   $previousValue \leftarrow myValue;$ $myValue \leftarrow argmin_{v \in D_{self}}$ LB(v);
 SendValueMessageToLowerNeighbors($myValue$);
 if (($self = root \wedge$ LB() $=$ UB()) $\vee$ ($end \wedge$ LB($myValue$) $=$ UB($myValue$))) **then**
   $end \leftarrow$ true;
   for each $child \in myChildren$ **do** sendMsg($STOP, self, child, false$);
 else
   $mySubtreeContr = myContribution$;
   for each $x_i \in myChildren$ **do** $mySubtreeContr = mySubtreeContr + childrenContribution[x_i]$;
   sendMsg($COST, self, myParent, myContext,$LB(), UB(), $mySubContribution$);

procedure CheckForRootDeletions()
 if $myPreviousValue \in D_{self} \wedge$ LB($myPreviousValue$) $>$UB() **then**
   $\top \leftarrow$ UB();   DeleteValue($myPreviousValue$);
 if LB($myValue$) $\geq$ UB() $\wedge$ UB($myValue$) $\neq$ UB() **then**
   $\top \leftarrow$ UB();   DeleteValue($myValue$);

procedure CheckForInternalDeletions()
 for each $val \in D(self)$ **do**
   $costValue = 0$;
   for each $child \in myChildren$ **do**
     if $tableContext[val, child].variable = self$ **then** $costValue = costValue + lb[val, child]$;
     else $costValue = costValue + childrenContribution[child]$;
   if $costValue > \top$ **then** DeleteValue(val);

procedure CheckCurrentContextWithChildren()
 for each $val \in D(self) \wedge child \in myChildren$ **do**
   if $\neg$isCompatible($myContext, tableContexts[val, child]$) **then**
     $tableContexts[val, child] \leftarrow$ empty; $lb[val, child] \leftarrow 0$; $ub[val, child] \leftarrow \infty$;

procedure SendValueMessageToLowerNeighbors($myValue$)
 $cost \leftarrow \sum_{j \in myParent \cup myPseudoparents} C_{self,j}(myValue, myContext[j])$;
 for each $child \in myChildren$ **do**
   $th \leftarrow min\{TH,$UB()$\} - cost - \sum_{j \in myChildren, j \neq child} lb[myValue, j]$;
   sendMsg($VALUE, self, child, myValue, th, \top, C_\phi$);

Fig. 4. BnB-ADOPT-AC* (cont.); missing procs appear in Fig 2

message is sent, *self* calculates the contribution of its subtree to the global C_ϕ accumulating its own contribution plus the contribution of its children and this is included in the COST message.

- CheckForRootDeletions. If $root = self$ and *myValue* or *myPreviousValue* are suboptimal they are deleted unconditionally and a new value is selected. Although discarded values from the *root* will never be revisited and there is no benefit in deleting them, the propagation of the DEL message may lead to further deletion.

- CheckForInternalDeletions. *self* calculates a lower bound for every domain value in such a way that its cost do not depend on any variable other than *self*. For this, it accumulates the lower bound informed by a child if the context of that child contains only the variable *self* (therefore the cost informed do not depend on any other higher assignation). Otherwise, it accumulates the *childrenContribution*, which is the minimum cost this child will have to pay for any higher variable assignation (is context free). If this calculated lower bound is greater than $\top$, we can remove this value unconditionally from the problem.

– SendValueMessageToLowerNeighbors. The $\top$ and C_ϕ values are included in the VALUE message.

4 Example

Consider variables x_1, x_2 and x_3 with domain $\{a, b\}$ and cost functions as represented in Figure 5 (left). Enforcing AC* on this problem (project on every variable costs from binary to unary constraints, and then project unary costs over a global C_ϕ) we get the equivalent problem of Figure 5 (right). As we have calculated the global C_ϕ, we already now that any solution will cost at least 17.

We present for this problem the execution trace of BnB-ADOPT (Table 1, left) and BnB-ADOPT-AC* (Table 1, right). As shown, BnB-ADOPT needs 42 messages and 9 cycles to reach the optimum solution, while BnB-ADOPT-AC* only needs 27 messages and 5 cycles. We will explain briefly why this happends. For a clearer explanation, we will omit some messages either because they are reiterative or because they are not relevant to show the benefits of BnB-ADOPT-AC*. For a more complete and detailed execution see Table 1.

First, all agents are initialized with value a and they send the correspondent VALUE messages (Figure 6(a)). Agent x_1 begins choosing value b, and receives the correspondent COST message (Figure 6(b)). Later on, x_1 tries value a, and receives the correspondent COST message (Figure 6(c)). Now, as all values of x_1 has been explored, x_1 chooses value b as best value for the current context, and sends a COST to the parent with an UB of 20 (Figure 6(d)). When x_0 receives this COST with UB different from ∞, it decides to change to value b and sends the correspondent VALUE messages to x_1 and x_2. When x_1 and x_2 receive the VALUE, they reinitialize their information (because context has changed), and they choose the value that minimize their LB under the

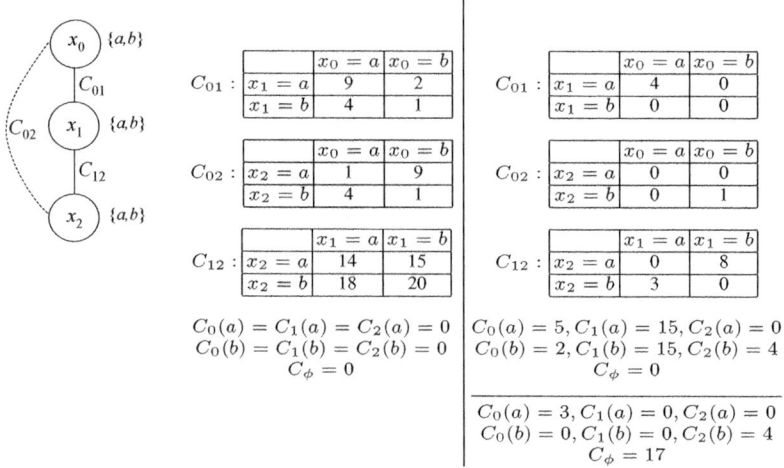

Fig. 5. (Left) Simple example with three variables and its inital binary, unary and zeroary cost functions. (Right) Above line: binary cost functions after proyection on unary ones; below line: unary cost functions after projecting on zeroary one.

current context; in this case x_1 chooses a again, and x_2 chooses b (Figure 6(e)). Agents x_2 and x_1 exchange messages until x_1 can send a COST message to its parent with an UB of 21 (Figure 6(f)). When x_0 receives this COST, as $UB(b) > UB(a)$, and all values has been explored, x_0 changes its value to a and terminates (Figure 6(g)). Then x_1 and x_2 exchange messages until they also terminate (Figures 6(h) and (i)). All states of Figure 6 indicate the line they represent on the execution trace.

In BnB-ADOPT-AC* execution, at Figure 6(d) (line 14 of execution trace), x_2 is capable of delete value b. This is because the global C_ϕ has been calculated and: $C_2(b) + C_\phi > \top$ (4+17 > 20). This deletion causes no inmediate effect, but is propagated to x_0. Then, when x_0 receives the DEL message (line 17 of execution trace) it reinforces AC* on itself, and as result of this x_0 can delete value b (line 18 of execution trace), since: $C_0(b) + C_\phi > \top$ (5+17 > 20). As $x_0 = b$ has been found not consistent and deleted, x_0 will not explore this value. So messages sent on Figure 6 (e) and (f) are not sent on BnB-ADOPT-AC* execution. Instead, x_0 sends a VALUE message with $x_0 = a$ and terminates (Figure 6 (g)). As we can see, deleting value b from x_0 has been beneficial, and the propagation of DEL messages has produced a positive impact.

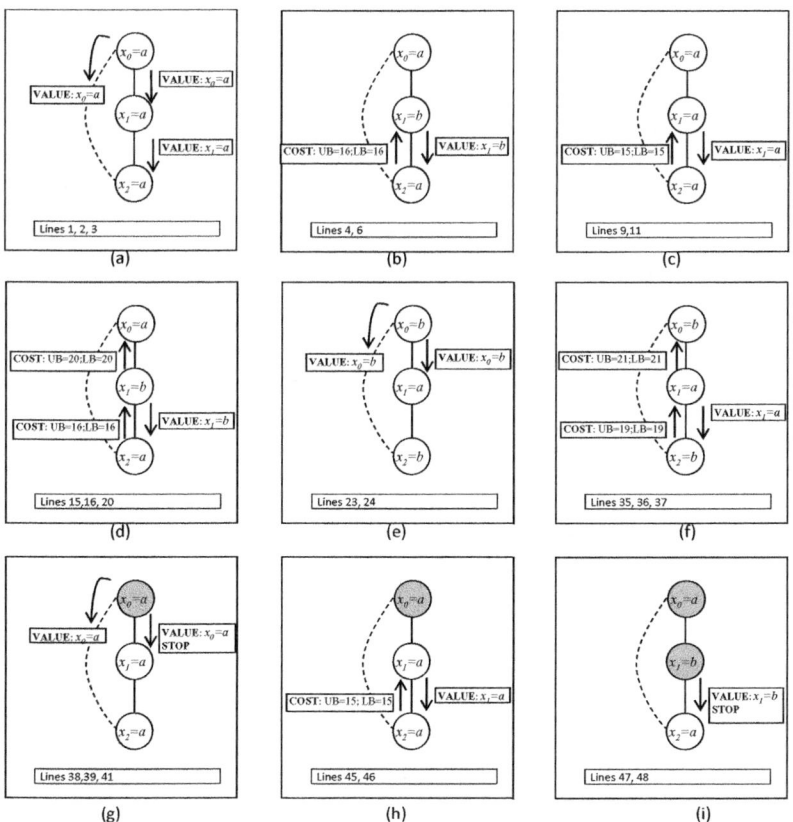

Fig. 6. Message passing example

Table 1. Trace of BnB-ADOPT and BnB-ADOPT-AC* on the example of Figure 5. Messages in bold are common to both algorithms. Notice that from lines 23 to 37 BnB-ADOPT messages are saved in BnB-ADOPT-AC*.

	BnB-ADOPT	BnB-ADOPT-AC*
(1)	x_1 **received VALUE:** $x_0 = a$	x_1 **received VALUE:** $x_0 = a$
(2)	x_2 **received VALUE:** $x_0 = a$	x_2 **received VALUE:** $x_0 = a$
(3)	x_2 **received VALUE:** $x_1 = a$	x_2 **received VALUE:** $x_1 = a$
(4)	x_2 **received VALUE:** $x_1 = b$	x_2 **received VALUE:** $x_1 = b$
(5)	x_0 **received COST: sender=**x_1**, UB=** ∞ **LB=4**	x_0 **received COST: sender=**x_1**, UB=** ∞**,LB=4 global** C_0**=17**
(6)	x_1 **received COST: sender=**x_2**, UB=16 LB=16**	x_1 **received COST: sender=**x_2**, UB=16, LB=16 global** C_0**=15**
(7)	x_1 **received VALUE:** $x_0 = a$	x_1 **received VALUE:** $x_0 = a$
(8)	x_2 **received VALUE:** $x_0 = a$	x_2 **received VALUE:** $x_0 = a$
(9)	x_2 **received VALUE:** $x_1 = a$	x_2 **received VALUE:** $x_1 = a$
(10)	x_0 **received COST: sender=**x_1**, UB=20 LB=9**	x_0 **received COST: sender=**x_1**, UB=20, LB=9 global** C_0**=17**
(11)	x_1 **received COST: sender=**x_2**, UB=15 LB=15**	x_1 **received COST: sender=**x_2**, UB=15, LB=15 global** C_0**=17**
(12)	x_1 **received VALUE:** $x_0 = a$	x_1 **received VALUE:** $x_0 = a$
(13)	x_2 **received VALUE:** $x_0 = a$	x_2 **received VALUE:** $x_0 = a$
(14)		x_2 delete value b
(15)	x_2 **received VALUE:** $x_1 = b$	x_2 **received VALUE:** $x_1 = b$
(16)	x_0 **received COST: sender=**x_1**, UB=20 LB=20**	x_0 **received COST: sender=**x_1**, UB=20, LB=20 global** C_0**=17**
(17)		x_0 received DEL: $x_2 = b$
(18)		x_0 delete value b
(19)		x_1 received DEL: $x_2 = b$
(20)	x_1 **received COST: sender=**x_2**, UB=16 LB=16**	x_1 **received COST: sender=**x_2**, UB=16, LB=16 global** C_0**=17**
(21)		x_1 delete value a
(22)		x_1 received DEL: $x_0 = b$
(23)	x_1 received VALUE: $x_0 = b$	
(24)	x_2 received VALUE: $x_0 = b$	
(25)	x_2 received VALUE: $x_1 = b$	
(26)	x_0 received COST: sender=x_1, UB= ∞ LB=1	
(27)	x_1 received COST: sender=x_2, UB=21 LB=21	
(28)	x_1 received VALUE: $x_0 = b$	
(29)	x_2 received VALUE: $x_0 = b$	
(30)	x_2 received VALUE: $x_1 = a$	
(31)	x_0 received COST: sender=x_1, UB=22 LB=2	
(32)	x_1 received COST: sender=x_2, UB=19 LB=19	
(33)	x_1 received VALUE: $x_0 = b$	
(34)	x_2 received VALUE: $x_0 = b$	
(35)	x_2 received VALUE: $x_1 = a$	
(36)	x_0 received COST: sender=x_1, UB=21 LB=21	
(37)	x_1 received COST: sender=x_2, UB=19 LB=19	
(38)	x_1 **received VALUE:** $x_0 = a$	x_1 **received VALUE:** $x_0 = a$
(39)	x_1 **received STOP**	x_1 **received STOP**
(40)		x_2 received DEL: $x_0 = b$
(41)	x_2 **received VALUE:** $x_0 = a$	x_2 **received VALUE:** $x_0 = a$
(42)	x_2 **received VALUE:** $x_1 = b$	
(43)		x_2 received DEL: $x_1 = a$
(44)	x_1 received COST: sender=x_2, UB=16 LB=16	
(45)	x_2 received VALUE: $x_1 = a$	
(46)	x_1 received COST: sender=x_2, UB=15 LB=15	
(47)	x_2 **received VALUE:** $x_1 = b$	x_2 **received VALUE:** $x_1 = b$
(48)	x_2 **received STOP**	x_2 **received STOP**
	No more messages...	No more messages...
	42 total messages	27 total messages
	24 VALUE msg , 16 COST msg	13 VALUE msg, 6 COST msg, 6 DEL msg
	9 cycles	5 cycles
	TOTAL cost: 20	TOTAL cost: 20
	OPT. SOLUTION: $x_0 = a$; $x_1 = b$; $x_2 = a$	OPT. SOLUTION: $x_0 = a$; $x_1 = b$; $x_2 = a$

5   Experimental Results

We evaluate the performance of BnB-ADOPT-AC* against BnB-ADOPT on binary random generated DCOPs and on meeting scheduling structured datasets [9].

The binary random DCOP sets are characterized by $\langle n, d, p_1 \rangle$, where n is the number of variables (#agents), d is the number of values per variable and p_1 the network connectivity defined as the ratio of existing cost functions. The cost of the tuples are selected from a uniform distribution of costs. Two types of binary cost functions are used, small and large. Small cost functions extract costs from the set $\{0, \ldots, 10\}$ while large ones extract costs from the set $\{0, \ldots, 1000\}$. The proportion of large cost functions is 1/4 of the total number of cost functions (this is done to introduce some variability among tuple costs; using a unique type of cost function causes that all tuples look pretty similar from an optimization view). We tested $\langle n = 10, d = 10, p_1 = 0.2, 0.3, 0.4, 0.5, 0.6, 0.7 \rangle$. Results of the execution appear in Table 2 (up), averaged over 50 instances.

The meeting scheduling structured datasets defines a DCOP equivalent to problems involving join events. On the presented formulation variables represent events, and each variable assigns a starting time for the event. Utility functions are constructed in such a way that when the DCOP is solved we obtain solutions congruent to the original problem. We present 4 cases with domain 10: case A (8 variables), case B (10 variables), case C (12 variables) and case D (12 variables). Results of the execution appear in Table 2 (down) , averaged over 30 instances.

Table 2 shows the number of messages exchanged (#VALUE, #COST, #DEL and the total number of messages), the number of cycles, the non-concurrent constraint checks and the average number of deletions in the domains of all variables. We evaluate the efficiency of the algorithm by a discrete event simulator. The total number of messages exchanged gives us a measure of the communication cost. The number of non-concurrent constraint checks [4] give us a measure of the computational effort needed to reach the solution. Finally, the cycles needed to solve the problem are the number of iterations that the simulator must perform until the solution is found. One cycle consist in every agent reading all incoming messages, performing local computation, and sending messages to neighbors.

BnB-ADOPT-AC* proved to be beneficial in the overall message passing for the connectivities tested, requiring less total messages than BnB-ADOPT (remember that it adds new DEL messages). The propagation of deletions contribute to diminish the search effort, decreasing the number of COST and VALUE messages exchanged. When the number of saved COST and VALUE messages is greater than the number of DEL messages, propagation pays off and causes an overall message decrement. We assume the usual case where the communication time is higher than computation time, then the total elapsed time is dominated by the communication time. In this case reducing the number of messages is beneficial, and also agents will need to process less information coming from their neighbors.

We also observe a clear decrement in the number of cycles required. The number of cycles often has been taken as a rough estimator of the efficiency of a distributed algorithm. We believe that it is a useful measure to compare algorithms that behave in a similar way. This is the case for BnB-ADOPT and BnB-ADOPT-AC*. Both send, per

Table 2. Results of BnB-ADOPT and BnB-ADOPT-AC* on: (up) random DCOP instances of 10 agents and 10 values per agent for $p_1 = 0.2, 0.3, 0.4, 0.5, 0.6, 0.7$; (down) Meting Scheduling, cases A (8 variables), B (10 variables), C (12 variables) and D (12 variables)

p_1	alg	#VALUE	#COST	#DEL	#Messages	#Cycles	#NCCC	#Del val
	BnB-ADOPT	346	333	0	688	39	**1,981**	0
0.2	BnB-ADOPT-AC*	264	233	143	**633**	**28**	9,347	77
	BnB-ADOPT	469,565	306,909	0	776,483	34,220	7,182,257	0
0.3	BnB-ADOPT-AC*	81,783	53,078	193	**135,064**	**5,916**	**804,565**	65
	BnB-ADOPT	9,583,678	4,902,157	0	14,485,844	547,459	132,524,817	0
0.4	BnB-ADOPT-AC*	1,969,416	996,734	211	**2,966,371**	**110,975**	**19,535,420**	56
	BnB-ADOPT	20,502,753	8,243,684	0	28,746,447	918,069	252,830,882	0
0.5	BnB-ADOPT-AC*	8,709,646	3,437,056	247	**12,146,959**	**382,322**	**97,612,718**	52
	BnB-ADOPT	17,617,714	6,134,986	0	23,752,710	685,054	189,687,091	0
0.6	BnB-ADOPT-AC*	8,381,955	2,836,996	298	**11,219,259**	**315,597**	**92,735,325**	53
	BnB-ADOPT	43,880,642	12,861,780	0	56,742,432	1,432,992	427,607,102	0
0.7	BnB-ADOPT-AC*	40,861,959	11,932,584	163	**52,794,716**	**1,329,283**	**394,671,491**	25

p_1	alg	#VALUE	#COST	#DEL	#Messages	#Cycles	#NCCC	#Del val
	BnB-ADOPT	66,641	29,845	0	96,493	4,427	697,774	0
A	BnB-ADOPT-AC*	20,381	8,999	177	**29,565**	**1,306**	**208,116**	43
	BnB-ADOPT	118,708	63,934	0	182,652	7,150	879,417	0
B	BnB-ADOPT-AC*	43,797	23,399	153	**67,358**	**2,615**	**303,634**	44
	BnB-ADOPT	20,664	13,698	0	34,374	1,278	167,058	0
C	BnB-ADOPT-AC*	5,351	3,490	208	**9,062**	**325**	**46,162**	73
	BnB-ADOPT	28,784	18,934	0	47,729	1,733	155,833	0
D	BnB-ADOPT-AC*	8,968	5,828	208	**15,017**	**533**	**51,755**	74

agent and per cycle, one COST message to the agent's parent and one VALUE message to each agent child and pseudochild. In addition, BnB-ADOPT-AC* sends DEL messages in some cycles. If BnB-ADOPT-AC* needs less cycles to find the optimum means that, in the same communication conditions, it will find the optimum faster than BnB-ADOPT. Finding the optimum means exhausting the search space: BnB-ADOPT-AC* does it more efficiently than BnB-ADOPT.

Computation effort measured as the number of non-concurrent constraint checks also decreases. This is the combination of two opposite trends: agents are doing more work processing new DEL messages but less work processing less VALUE and COST messages. The overall picture indicates that adding DEL messages makes smaller agent domains, so it reduces the search space. To explore this reduced search space in a distributed way, less VALUE and COST messages are needed. This decrement is higher than DEL messages increment.

6   Conclusions

In this work we have connected the BnB-ADOPT algorithm with some forms of soft arc consistency (weighted case) aiming at detecting and pruning values which would not be in the optimal solution, with the final goal of improving search efficiency. These deletions are unconditional and do not rely on any previous variable assignment. The transformations introduced (projecting costs from binary to unary, from unary to zero-ary, and pruning values not NC*) assure that the optimum of the transformed problem remains the same as the original one.

According to experimental results, propagation of unconditional deletions provides substantial benefits for the sets tested. A new message DEL has been introduced for the propagation of deleted values. However, the increment in the number of messages due to the generation of new DEL messages has been compensated by a decrement in the number of COST and VALUE messages used to solve the problem. BnB-ADOPT-AC* has been proved to be beneficial regarding the total number of messages exchanged, the number of non-concurrent constraint checks performed and the number of cycles required to find the optimum.

As future work, we consider the use of other soft local consistencies such as DAC* or FDAC* [3]. We also consider the application of these techniques in other ADOPT-based algorithms like ADOPT-ng [7].

References

1. Dechter, R.: Constraint Processing. Morgan Kaufmann, San Francisco (2003)
2. Larrosa, J.: Node and arc consistency in weighted CSP. In: Proc. of AAAI 2002 (2002)
3. Larrosa, J., Schiex, T.: In the quest of the best form of local consistency for weighted CSP. In: Proc. of IJCAI 2003 (2003)
4. Meisels, A., Kaplansky, E., Razgon, I., Zivan, R.: Comparing performance of distributed constraint processing algorithms. In: AAMAS Workshop on Distributed Constraint Reasoning, pp. 86–93 (2002)
5. Meseguer, P., Rossi, F., Schiex, T.: Handbook of Constraint Programming. In: Soft Constraints, ch. 9. Elsevier, Amsterdam (2006)
6. Modi, P.J., Shen, W.M., Tambe, M., Yokoo, M.: Adopt: asynchronous distributed constraint optimization with quality guarantees. Artificial Intelligence (161), 149–180 (2005)
7. Silaghi, M., Yokoo, M.: Nogood-based asynchronous distributed optimization (ADOPT-ng). In: Proc. of AAMAS 2006 (2006)
8. Yeoh, W., Felner, A., Koenig, S.: Bnb-adopt: An asynchronous branch-and-bound DCOP algorithm. In: Proc. of AAMAS 2008, pp. 591–598 (2008)
9. Yin, Z.: USC dcop repository. Meeting scheduling and sensor net datasets (2008), http://teamcore.usc.edu/dcop

Procedural Code Generation *vs* Static Expansion in Modelling Languages for Constraint Programming

Julien Martin, Thierry Martinez, and François Fages

EPI Contraintes, INRIA Paris-Rocquencourt,
BP105, 78153 Le Chesnay Cedex, France
{Julien.Martin,Thierry.Martinez,Francois.Fages}@inria.fr
http://contraintes.inria.fr/

Abstract. To make constraint programming easier to use by the non-programmers, a lot of work has been devoted to the design of front-end modelling languages using logical and algebraic notations instead of programming constructs. The transformation to an executable constraint program can be performed by fundamentally two compilation schemas: either by a static expansion of the model in a flat constraint satisfaction problem (e.g. Zinc, Rules2CP, Essence) or by generation of procedural code (e.g. OPL, Comet). In this paper, we compare both compilation schemas. For this, we consider the rule-based modelling language Rules2CP with its static exansion mechanism and describe with a formal system a new compilation schema which proceeds by generation of procedural code. We analyze the complexity of both compilation schemas, and present some performance figures of both the compilation process and the generated code on a benchmark of scheduling and bin packing problems.

1 Introduction

Constraint programming is a programming paradigm which relies on two components: a constraint component which manages posting and checking satisfiability and entailment of constraints over some fixed computational domain, and a programming component which makes it possible to state the constraints of a given problem and define a search procedure for solving it. To make constraint programming easier to use by non-programmers, a lot of work has been devoted to the design of front-end modelling languages using logical and algebraic notations instead of programming constructs, e.g. OPL[14,7], Comet [10] , Zinc [11,3], Essence [6] or Rules2CP [4,5,2].

Such modelling languages for constraint programming offer a high-level of abstraction for stating constraint problems, and rely on default, possibly parameterized or adaptive, search strategies. The transformation to an executable constraint program can be performed by fundamentally two compilation schemas: either by a static expansion of the model in a flat constraint satisfaction problem, or by generation of procedural code. The first schema by static expansion

J. Larrosa and B. O'Sullivan (Eds.): CSCLP 2009, LNAI 6384, pp. 38–58, 2011.

has been adopted by Zinc, Essence and Rules2CP, while the second schema by code generation has been implemented for OPL and Comet.

In this paper, we compare both compilation schemas. For this, we consider the rule-based modelling language Rules2CP with its static exansion mechanism described in [4], and introduce a new compilation schema which proceeds by generation of procedural code. With this new implementation, called Cream, we show that the code generation schema exhibits a time overhead of approximatively a factor 2 at runtime w.r.t. the statically expanded code. However, we show that the size of the procedural code is linear, which must be compared to the potentially exponential size of the expanded code. In particular, for problems where the search space is defined dynamically by values of variables at runtime, the code generation schema is the only viable one.

Furthermore, in a rule-based modelling language such as Rules2CP, the search tree is represented by a logical formula and search tree ordering heuristics can be expressed declaratively by pattern-matching on the rules' left-hand sides [5]. Compared with other modelling languages capable of expressing search heuristics, such as OPL/Comet for instance, rule-based pattern matching eliminates the need to program with lists and indices and to introduce data structures for defining the ordering criteria. Compared with Zinc, this mechanism provides a possible mean to define heuristics for the default search procedure. The price to pay for this expressivity however is in the compilation process which becomes more complicated. This was our original motivation for defining the transformations with a formal system.

The rest of the paper is organized as follows. The next section defines the syntax of Rules2CP, its polymorphic type system and the declarative semantics of the language. Section 3 defines the static expansion schema with a formal system that is reused in section 4 to define the code generation schema and to analyze their complexity. Section 5 evaluates the performance of both compilation schema and generated code on a benchmark of n-queens, scheduling and bin packing problems. Finally we conclude on the merits of each compilation schema.

2   Rules2CP Syntax and Declarative Semantics

2.1   Syntax

There are four data structures in Rules2CP:

- integer constants, with basic arithmetic operators and comparisons.
- finite domain variables, with indexicals and the equality constraint in addition to the operators shared with integer constants.
- lists, constructed by enumeration, interval between two integers and concatenation, and browsed with quantifiers and aggregators.
- records, with labeled fields used for projection.

The Rules2CP syntax is summarized in table 1. The non-terminal *variable* and *ident* range over a countable set of names. The non-terminal *integer* ranges over a finite interval $\mathcal{D} \subseteq \mathbf{N}$ which includes at least the values 0 and 1. Underlined non-terminal *var* mark the binders which affect the underlined *expr*.

The sets of bound and free variables in an expression e, denoted bv(e) and fv(e) respectively, are defined in the standard way: a variable is bound if it is in the scope of a binder (let or foldl) or if it appears in the left-hand side of a clause. Any assignment $\nu : var \rightarrow \mathcal{D}$ is homomorphically extended to a function $\tilde{\nu} : expr \rightarrow expr$.

Table 1. Rules2CP syntax

$$
\begin{aligned}
program &::= clause \; ... \; clause \\
clause &::= \texttt{domain } ident \; := \; \{ \; ident, ..., ident \; \} \\
&\mid \; \texttt{object } ident(\underline{var}, ..., \underline{var}) \; := \; expr \\
&\mid \; \texttt{rule } ident(\underline{var}, ..., \underline{var}) \; := \; expr \\
&\mid \; \texttt{heuristics } ident(\underline{var}, ..., \underline{var}) \; := \; heuristics \\
&\mid \; \texttt{query } expr \\
expr &::= variable \mid integer \mid \texttt{error} \\
&\mid \; expr \; op \; expr \; \text{where } op \in \{\texttt{+}, \texttt{-}, \texttt{*}, \texttt{/}\} \\
&\mid \; expr \; rel \; expr \; \text{where } rel \in \{\texttt{=}, \texttt{\#}, \texttt{=<}, \texttt{<}, \texttt{>}, \texttt{>=}\} \\
&\mid \; expr \; logop \; expr \; \text{where } logop \in \{\texttt{and}, \texttt{or}, \texttt{implies}, \texttt{equiv}\} \\
&\mid \; \texttt{not } expr \\
&\mid \; ident(expr, ..., expr) \\
&\mid \; \texttt{let}(\underline{var} \; := \; expr, \; expr) \\
&\mid \; [expr, ..., expr] \mid [\underline{expr} \; .. \; expr] \mid expr \; \texttt{++} \; expr \\
&\mid \; \texttt{length}(expr) \mid \texttt{nth}(expr, \; expr) \\
&\mid \; \{ident: \; expr, ..., ident: \; expr\} \mid expr:ident \\
&\mid \; \texttt{foldl}(\underline{var} \; \texttt{from} \; expr, \underline{var} \; \texttt{in} \; expr, expr) \\
&\mid \; \texttt{minimize}(expr, expr) \mid \texttt{maximize}(expr, expr) \\
&\mid \; \texttt{search}(heuristics, expr) \mid \texttt{constraint}(expr) \\
&\mid \; \texttt{dynamic}(expr) \mid \texttt{static}(expr) \\
heuristics &::= \texttt{conjunctive}(\underline{expr} \; \texttt{for} \; ident(\underline{var}, ..., \underline{var})) \\
&\mid \; \texttt{disjunctive}(\underline{expr} \; \texttt{for} \; ident(\underline{var}, ..., \underline{var})) \\
&\mid \; ident(expr, ..., expr) \\
&\mid \; \texttt{nil} \mid heuristics \; \texttt{and} \; heuristics
\end{aligned}
$$

Free variables are not allowed in rule definitions. Free variables in object definitions are allowed and denote finite-domain variables. They are indexed by the head of the definition. For instance, the following definition introduces a new finite-domain variable in the field `row` for each value of I.

```
object queen(I)  := { row : _, column : I }.
```

The concrete Rules2CP implementation introduces some syntactic sugar:

- the let-construction is recursively extended for multiple bindings. For all n, the let of $n+1$ bindings $\texttt{let}(X_0 := e_0, ..., X_n := e_n, e)$ is defined with the simple let and the let of n bindings: $\texttt{let}(X_0 := e_0, \texttt{let}(X_1 := e_1, ..., X_n := e_n, e))$.
- $\texttt{forall}(X \; \texttt{in} \; l, \; e)$ is a synonym for $\texttt{foldl}(A \; \texttt{from} \; 1, \; X \; \texttt{in} \; l, \; A \; \texttt{and} \; e)$ where A is a fresh variable.

- exists(X in l, e) is foldl(A from 0, X in l, A or e) where A is a fresh variable.
- map(X in l, e) is foldl(A from [], X in l, A ++ [e]) where A is a fresh variable.
- reverse(l) is foldl(A from [], X in l, [X] ++ A) where A and X are fresh variables.
- foldr(A from i, X in l, e) is foldl(A from i, X in reverse(l), e) where A is a fresh variable.

Example 1. The classical (unavoidable) n-queens problem can be modelled in Rules2CP as follows. First, the board of queens can be defined by the following object definitions:

```
object queen(I) := { row : _, column : I }.
object board(N) := map(I in [1 .. N], queen(I)).
```

For each integer I, queen(I) defines one record representing the queen in column I. Then, the goal is to post the constraints over the list of queens B and assign values to the free variables.

```
? let(N := 4, B := board(N),
    queens_constraints(B, N) and
    queens_labeling(variables(B), N)).
```

The predicate queens_constraints is defined by the following rules.

```
rule queens_constraints(B, N) := domain(B, 1, N) and safe(B).
```

```
rule safe(L) :=
  all_different(L) and
  forall(Q in L, forall(R in L,
    let(I := Q:column, J := R:column,
      I < J implies
      Q:row # J - I + R:row and
      Q:row # I - J + R:row))).
```

The rule safe(L) ensures that every queen in the list L is on a safe position: the global constraint all_different prevents row attacks and simple binary difference constraints prevents diagonal attacks.

The rule for queens_labeling(Vars, N) defines the search through a logical formula which induces a basic labeling search tree on variables Vars.

```
rule queens_labeling(Vars, N) :=
  search(mo(N), forall(Var in Vars, queens_labeling_var(Var, N))).
```

```
rule queens_labeling_var(Var) :=
  exists(Val in [1 .. N], queens_labeling_val(Var, Val)).
```

```
rule queens_labeling_val(Var, Val) := Var = Val.
```

42   J. Martin, T. Martinez, and F. Fages

The parameter mo(N) for search refers to the *middle out* heuristics which is
defined by the following rule:

```
heuristics mo(N) :=
  disjunctive(least(abs(N/2 - Val))
                   for queens_labeling_val(Var, Val)).
```

This statement specifies that the disjunctive formulae derived from
queens_labeling_val must be ordered by increasing value of abs(N/2 - Val)
(middle out ordering of values).

2.2 Type System

Rules2CP integrates a type system with five type constructors:

- int for integer values.
- fd for finite domain variables.
- constraint for first-order logic formulas.
- $[\tau]$ for (homogeneous) lists whose elements have type τ.
- $\{f_1 :: \tau_1, \ldots, f_n :: \tau_n\}$ for records with the fields $f_1, \ldots, f_n$ carrying
  values of type $\tau_1, \ldots, \tau_n$ respectively.

A free variable in a Rules2CP program is always an FD variable. The boolean
values true and false are not distinguished from the integers 1 and 0.

The type system enjoys a type inference algorithm *à la* Hindley-Milner: typing
rules are driven by the syntax of the expression and induce type equality constraints solved by unification. Type schemes with universal quantification are
given to polymorphic definitions and rules where arguments are not completely
specified.

Arithmetic operators and comparisons are overloaded to deal with both integer values and FD variables. For example, the addition operator is typed
int + int :: int if both arguments are known to be of type int, otherwise it
is typed fd + fd :: fd. It is worth noting that Hindley-Milner does not allow
ad-hoc overloading in general. Here we made the choice to use int to follow
statically known integer values and fd for model variables. Some constructions
are specific to integer values: in particular, the list interval constructor has type
[int .. int] :: [int]. Indexical built-ins transform FD variables into integer
values: min(fd) :: int and max(fd) :: int.

Records are typed with *row types* [12], and two records are equal when they
have the same set of fields and when fields of the same name carry values of
equal types.

Example 2. Let us consider the two following rules defining the area and volume
of an object. The argument X is only accessed by projection and can be of
any record type containing at least the fields width and height (and depth for
volume).

```
area(X) = X:width * X:height
volume(X) = area(X) * X:depth
```

In the inferred type,

```
area({ height: fd, width: fd, A }) :: fd
volume({ depth: fd, height: fd, width: fd, A }) :: fd
```

the unknown other fields are symbolized by a *row variable*. Such a row type containing a row variable is said to be open. Row types without row variables are closed.

In the following `shape` object definition

```
shape(Id) = { id: Id, width = _, height = _ }
```

the parameter `Id` can be of any type. The type inferred for `shape` is polymorphic and parameterized by a *type variable* `A` given to the argument `Id`. The other arguments are model variables (free variables at the right-hand side of a definition) and are therefore typed with `fd`.

```
shape(A) :: { id: A, width: fd, height: fd }
```

The Hindley-Milner type inference with row types is known to be decidable with a theoretical PSPACE-hard time complexity [8]. However, this worst-case time complexity does not exhibit in practice and the type inference algorithm is very efficient.

2.3   Declarative Semantics

Let $\mathcal{M}$ be a Rules2CP model. Let $O(\mathcal{M})$ be the set of all the objects of $\mathcal{M}$, let $R(\mathcal{M})$ be the set of all the rules, and $Q(\mathcal{M})$ be the set of all the queries of $\mathcal{M}$. Queries are interpreted conjunctively: the query associated to $\mathcal{M}$ is $q(\mathcal{M}) = \bigwedge_{q \in Q(\mathcal{M})} q$.

This section will characterize the solutions of the Rules2CP model $\mathcal{M}$. A solution is an assignment of all the free variables of $\mathcal{M}$ which satisfies all the constraints of $\mathcal{M}$. Free variables occurs in the query $q(\mathcal{M})$ and in object definitions. The free variables in object definitions are distinct for each instance of the object. The arguments of an object are restricted to belong to the following grammar.

$$indexable ::= integer$$
$$| \quad [indexable, ..., indexable]$$
$$| \quad \{ident: expr, ..., ident: expr\}_{uid}$$

Each indexable value v defines an index $id(v)$ which serves to index the free variables appearing in the object definition.

$$id: \qquad\qquad indexable \rightarrow index$$
$$i \in integer \mapsto \mathtt{constant}(i)$$
$$[i_1, ..., i_n] \mapsto [id(i_1), ..., id(i_n)]$$
$$\{ident: expr, ..., ident: expr\}_{uid} \mapsto \mathtt{uid}(uid)$$

An assignment for $\mathcal{M}$ is a tuple (ν^Q, ν^O), where:

- $\nu^Q : \mathrm{fv}(q(\mathcal{M})) \to \mathcal{D}$
- ν^O is a family of assignments which maps every object $o \in O(\mathcal{M})$ and every tuple $(i_1, \ldots, i_n) \in \mathit{index}^n$, where n is the arity of the head of d, to an assignment $\nu^O_{o(i_1,\ldots,i_n)} : \mathrm{fv}(o) \to \mathcal{D}$

Table 2. Small-step reduction semantics defining the success semantics of Rules2CP (without distinguishing optimization from satisfaction predicates)

$$n \in \mathbf{N} \;\; op \;\; n' \in \mathbf{N} \to n \; op \; n'$$

$$n \in \mathbf{N} \;\; rel \;\; n' \in \mathbf{N} \to \delta(n \; rel \; n')$$

$$n \in \{0,1\} \;\; logop \;\; n' \in \{0,1\} \to \delta(n = 1 \; logop \; n' = 1)$$

$$\mathbf{not} \;\; n \in \{0,1\} \to \delta(n = 0)$$

$$e \to \tilde{\nu}^Q(e)$$
$$\text{if } \mathbf{query} \;\; e \in Q(\mathcal{M})$$

$$o(e_1,\ldots,e_n) \to \tilde{\nu}^O_{o(id(e_1),\ldots,id(e_n))}(e)[X_1 := e_1, \ldots, X_n := e_n]$$
$$\text{if } d = \mathbf{object} \;\; o(X_1,\ldots,X_n) \;\; := \; e \in O(\mathcal{M})$$
$$\text{and } (e_1,\ldots,e_n) \in \mathit{indexable}^n$$

$$p(e_1,\ldots,e_n) \to e[X_1 := e_1, \ldots, X_n := e_n]$$
$$\text{if } d = \mathbf{rule} \;\; p(X_1,\ldots,X_n) \;\; := \; e \in R(\mathcal{M})$$

$$\mathbf{let}(x := v, e) \to e[x := v]$$

$$[n \in \mathbf{N} \;\; .. \;\; n' \in \mathbf{N}] \to \begin{cases} [n, \; n+1, \ldots, n'] & \text{if } n \le n' \\ [] & \text{otherwise} \end{cases}$$

$$[e_1,\ldots,e_n] \;\; \mathbf{++} \;\; [e_1',\ldots,e_n'] \to [e_1,\ldots,e_n, e_1',\ldots,e_n']$$

$$\mathbf{length}([e_1, \ldots, e_n]) \to n$$

$$\mathbf{nth}(i \in \{1,\ldots,n\}, \; [e_1, \ldots, e_n]) \to e_i$$

$$\{f_1 : e_1,\ldots, f_n : e_n\} : f_i \to e_i$$

$$\mathbf{foldl}(A \; \mathbf{from} \; i, X \; \mathbf{in} \; [e_1,\ldots,e_n], e) \to i \rhd_e e_1 \rhd_e \cdots \rhd_e e_n$$
$$\text{where } u \rhd_e v = e[A := u, X := v]$$

$$\left. \begin{array}{r} \mathbf{minimize}(g,k) \\ \mathbf{maximize}(g,k) \\ \mathbf{search}(h,g) \\ \mathbf{constraint}(g) \\ \mathbf{static}(g) \\ \mathbf{dynamic}(g) \end{array} \right\} \to g$$

Let δ be the reification operator: $\delta(\top) = 1$ and $\delta(\bot) = 0$. A solution for a model $\mathcal{M}$ is an assignment (ν^Q, ν^O) for which the query of $\mathcal{M}$ is reduced to 1 by the small-step reduction described in table 2.

Definition 1. *The set of observables $\mathcal{O}_s(\mathcal{M})$ for the success semantics of $\mathcal{M}$ is the set of solutions of $\mathcal{M}$.*

$$\mathcal{O}_s(\mathcal{M}) = \{(\nu^Q, \nu^O) \mid \nu^Q(q(\mathcal{M})) \xrightarrow{*} 1\}$$

3 Static Expansion Schema

The static expansion schema is defined by two transformations, the first one producing intermediate code:

1. $—\langle\text{stc}\rangle\longrightarrow$ expands a query to the deterministic code which adds the constraints
2. $—\langle{}^{\text{stc}}_{\text{srch}}\rangle\longrightarrow$ expands the search code.

The elimination of negations in formulae by descending them to the constraints with De Morgans laws are part of transformations, but are not presented.

3.1 Deterministic Code Generation

Built-in operators. Reification transforms boolean values in integers and logical operators in artihmetic operators. Partial evaluation occurs on arithmetic, comparison and logical operators.

$$\frac{e_1 —\langle\text{stc}\rangle\longrightarrow e_1' \qquad e_2 —\langle\text{stc}\rangle\longrightarrow e_2'}{e_1 \; op \; e_2 —\langle\text{stc}\rangle\longrightarrow e_1' \; op \; e_2'}$$

$$\frac{e_1 —\langle\text{stc}\rangle\longrightarrow e_1' \qquad e_2 —\langle\text{stc}\rangle\longrightarrow e_2'}{e_1 \; rel \; e_2 —\langle\text{stc}\rangle\longrightarrow \texttt{reify}(e_1' \; rel \; e_2')}$$

$$\frac{e_1 —\langle\text{stc}\rangle\longrightarrow e_1' \qquad e_2 —\langle\text{stc}\rangle\longrightarrow e_2'}{e_1 \; logop \; e_2 —\langle\text{stc}\rangle\longrightarrow \texttt{reify}(e_1' = 1 \; logop \; e_2' = 1)}$$

$$\frac{e —\langle\text{stc}\rangle\longrightarrow e'}{\texttt{not} \; e —\langle\text{stc}\rangle\longrightarrow \texttt{reify}(e' = 0)}$$

Definitions and calls. Within this static expansion schema, definitions are fully expanded. Free variables in object definitions are indexed and stored in a table ν^O.

$$a_1 —\langle\text{stc}\rangle\longrightarrow a_1'$$

$$\cdots$$

$$\frac{a_n —\langle\text{stc}\rangle\longrightarrow a_n' \qquad e[X_1 := a_1', \ldots, X_n := a_n'] —\langle\text{stc}\rangle\longrightarrow e'}{p(a_1, \ldots, a_n) —\langle\text{stc}\rangle\longrightarrow e'} \quad \begin{cases} r = \texttt{rule} \; p(X_1, \ldots, X_n) := e \in R(\mathcal{M}) \\ \texttt{fv}(r) = \emptyset \end{cases}$$

$$\frac{\begin{array}{c} a_1 \;—\langle\text{stc}\rangle\!\!\longrightarrow a_1' \\ \cdots \\ a_n \;—\langle\text{stc}\rangle\!\!\longrightarrow a_n' \\ \sigma(e)[X_1 := a_1', \ldots, X_n := a_n'] \;—\langle\text{stc}\rangle\!\!\longrightarrow e' \end{array}}{p(a_1, \ldots, a_n) \;—\langle\text{stc}\rangle\!\!\longrightarrow e'} \quad \begin{cases} d = \textbf{object } o(X_1, \ldots, X_n) := e \in R(\mathcal{M}) \\ \sigma = \nu_{o(id(a_1'), \ldots, id(a_n'))}^O \\ \text{dom}(\sigma) = \text{fv}(d) \end{cases}$$

Lists. If its bounds are statically instantiated, a range is reduced to the list of integers that it contains by partial evaluation.

$$\frac{e_1 \;—\langle\text{stc}\rangle\!\!\longrightarrow l \quad \cdots \quad e_n \;—\langle\text{stc}\rangle\!\!\longrightarrow u}{[e_1 \;\; .. \;\; e_2] \;—\langle\text{stc}\rangle\!\!\longrightarrow [l, \; l+1, \ldots, \; u]} \quad \begin{cases} l, u \in \mathbf{N} \\ l \le u \end{cases}$$

$$\frac{l_1 \;—\langle\text{stc}\rangle\!\!\longrightarrow [d_1, \ldots, d_n] \quad l_2 \;—\langle\text{stc}\rangle\!\!\longrightarrow [e_1, \ldots, e_m]}{l_1 \;\texttt{++}\; l_2 \;—\langle\text{stc}\rangle\!\!\longrightarrow [d_1, \ldots, d_n, \; e_1, \ldots, e_n]}$$

$$\frac{e_1 \;—\langle\text{stc}\rangle\!\!\longrightarrow e_1' \quad \cdots \quad e_n \;—\langle\text{stc}\rangle\!\!\longrightarrow e_n'}{[e_1, \ldots, e_n] \;—\langle\text{stc}\rangle\!\!\longrightarrow [e_1', \ldots, e_n']}$$

Records. Record projection need the record to be statically instantiated.

$$\frac{e_i \;—\langle\text{stc}\rangle\!\!\longrightarrow e_i'}{\{f_1\colon e_1, \ldots, f_n\colon e_n \}\colon f_i \;—\langle\text{stc}\rangle\!\!\longrightarrow e_i'} \quad f_i \in \{f_1, \ldots, f_n\}$$

$$\frac{e_1 \;—\langle\text{stc}\rangle\!\!\longrightarrow e_1' \quad \cdots \quad e_n \;—\langle\text{stc}\rangle\!\!\longrightarrow e_n'}{\{f_1\colon e_1, \ldots, f_n\colon e_n \} \;—\langle\text{stc}\rangle\!\!\longrightarrow \{f_1\colon e_1', \ldots, f_n\colon e_n' \}}$$

Let-binding. Substitutions are implicitly operated modulo alpha-conversion.

$$\frac{v \;—\langle\text{stc}\rangle\!\!\longrightarrow v'}{\texttt{let}(X := v, \; e) \;—\langle\text{stc}\rangle\!\!\longrightarrow e[X := v']}$$

Combinators. Combinators are expanded and require their list and initial element arguments to be statically instantiated.

$$\frac{\begin{array}{c} i \;—\langle\text{stc}\rangle\!\!\longrightarrow i_0 \quad l \;—\langle\text{stc}\rangle\!\!\longrightarrow [e_1, \ldots, e_n] \\ i_0 \triangleright_e e_1 \;—\langle\text{stc}\rangle\!\!\longrightarrow i_1 \\ i_1 \triangleright_e e_2 \;—\langle\text{stc}\rangle\!\!\longrightarrow i_2 \quad \cdots \quad i_{n-1} \triangleright_e e_n \;—\langle\text{stc}\rangle\!\!\longrightarrow i_n \end{array}}{\texttt{foldl}(A \text{ from } i, \; X \text{ in } l, \; e) \;—\langle\text{stc}\rangle\!\!\longrightarrow i_n}$$

where $u \triangleright_e v = e[A := u, X := v]$.

Search. By default, a logic formula f defines a reified constraint. In the context of a $\texttt{search}(f)$ predicate, f defines a search tree.

$$\frac{f \xrightarrow{\;\langle\,^{\text{stc}}_{\text{srch}}\rangle\;} f'}{\texttt{search}(f) \xrightarrow{\;\langle\text{stc}\rangle\;} f'} \qquad\qquad \frac{f \xrightarrow{\;\langle\text{stc}\rangle\;} f'}{\texttt{constraint}(f) \xrightarrow{\;\langle\text{stc}\rangle\;} f'}$$

A predicate $\texttt{minimize}(f,\ c)$ minimizes the value of the finite domaine variable V denoted by c following a branch and bound search. f is a formula implicitly interpreted as a search tree that constrain V to an assignment.

$$\frac{\texttt{search}(e) \xrightarrow{\;\langle\,^{\text{stc}}_{\text{srch}}\rangle\;} e' \qquad c \xrightarrow{\;\langle\text{stc}\rangle\;} V}{p(e,\ c) \xrightarrow{\;\langle\text{stc}\rangle\;} p(e',\ V)} \quad \begin{cases} p \in \{\texttt{minimize}, \\ \qquad\quad \texttt{maximize}\} \end{cases}$$

Dynamic mode. It is possible to dynamically evaluate (see Sec. 4) an expression instead of statically expand it with the predicate $\texttt{dynamic/1}$.

$$\frac{e \xrightarrow{\;\langle\text{dyn}\rangle\;} e'}{\texttt{dynamic}(e) \xrightarrow{\;\langle\text{stc}\rangle\;} e'} \qquad\qquad \frac{e \xrightarrow{\;\langle\text{stc}\rangle\;} e'}{\texttt{static}(e) \xrightarrow{\;\langle\text{stc}\rangle\;} e'}$$

Example 3. For the n-queens model presented in example 1, the static expansion compilation schema procudes the following intermediate code for $n = 4$:

```
domain([Q_1_1,Q_2_1,Q_3_1,Q_4_1], 1, 4) and
all_different([Q_1_1,Q_2_1,Q_3_1,Q_4_1]) and
Q_1_1 # 1+Q_2_1 and Q_1_1 # -1+Q_2_1 and
Q_1_1 # 2+Q_3_1 and Q_1_1 # -2+Q_3_1 and
Q_1_1 # 3+Q_4_1 and Q_1_1 # -3+Q_4_1 and
Q_2_1 # 1+Q_3_1 and Q_2_1 # -1+Q_3_1 and
Q_2_1 # 2+Q_4_1 and Q_2_1 # -2+Q_4_1 and
Q_3_1 # 1+Q_4_1 and Q_3_1 # -1+Q_4_1 and

search(Q_1_1 = 2 or Q_1_1 = 3 or Q_1_1 = 1 or Q_1_1 = 4 and
       Q_2_1 = 2 or Q_2_1 = 3 or Q_2_1 = 1 or Q_2_1 = 4 and
       Q_3_1 = 2 or Q_3_1 = 3 or Q_3_1 = 1 or Q_3_1 = 4 and
       Q_4_1 = 2 or Q_4_1 = 3 or Q_4_1 = 1 or Q_4_1 = 4)
```

It is worth noting that the complete cartesian product of all queens is not generated for the binary difference constraints thanks to the partial evaluation mechanism.

3.2   Non-deterministic Code Generation

By the $\xrightarrow{\;\langle\,^{\text{stc}}_{\text{srch}}\rangle\;}$ transformation, the conjunction operator and becomes a sequence operator and the disjunction operator or becomes a non-deterministic choice operator.

The formula f is expanded following $\xrightarrow{\;\langle\text{stc}\rangle\;}$ schema and the modification described above is operated giving a non-deterministic code $f' : f \xrightarrow{\;\langle\,^{\text{stc}}_{\text{srch}}\rangle\;} f'$.

3.3 Correctness and Complexity of the Static Expansion Schema

Proposition 1. *Given a Rule2CP model $\mathcal{M}$, let $\mathcal{M}'$ such that $\mathcal{M} \longrightarrow\!\langle\text{stc}\rangle\!\longrightarrow \mathcal{M}'$, then $\mathcal{O}_s(\mathcal{M}) = \mathcal{O}_s(\mathcal{M}')$ (i.e., $\longrightarrow\!\langle\text{stc}\rangle\!\longrightarrow$ preserves the model declarative semantics.)*

Proof. For every assignment (ν^Q, ν^O) for $\mathcal{M}$, we check inductively on the derivation of $\longrightarrow\!\langle\text{stc}\rangle\!\longrightarrow$ and $\longrightarrow\!\langle{}^{\text{stc}}_{\text{srch}}\rangle\!\longrightarrow$ that $\nu^Q(\mathcal{M}) \xrightarrow{*} \nu^Q(\mathcal{M}')$. Most of derivations are independent from assignment and verify this property by definition. Calls to object definitions are restricted to indexable arguments and the table ν^O is used for indexation. $\longrightarrow\!\langle{}^{\text{stc}}_{\text{srch}}\rangle\!\longrightarrow$ schema does not change the set of solutions with respect to $\longrightarrow\!\langle\text{stc}\rangle\!\longrightarrow$.

Definition 2. *Given a Rule2CP model $\mathcal{M}$, the* fold rank *$\alpha(s)$ of a symbol s is defined inductively by:*

$\alpha(s) = 0$ *if s is not the head symbol of a declaration or rule in $\mathcal{M}$,*
$\alpha(s) = max\{n + \alpha(s') \mid L = R \in \mathcal{M}$, s is the head symbol of L and R contains a nesting of n fold operators or quantifiers on an expression containing symbol $s'\}$.

The fold rank of $\mathcal{M}$ is the maximum fold rank of the symbols in $\mathcal{M}$.

Definition 3. *the* definition rank *$\rho(s)$ of a symbol s is defined inductively by:*

$\rho(s) = 0$ *if s is not the head symbol of a clause in $\mathcal{M}$,*
$\rho(s) = n + 1$, *if s is the head symbol of a clause in $\mathcal{M}$ and n is the greatest definition rank of the symbols in the right hand side of the clause.*

The definition rank of $\mathcal{M}$ is the maximum definition rank of the symbols defined in $\mathcal{M}$.

Proposition 2. *[4] For any* Rules2CP *model $\mathcal{M}$, the size of the generated program is in $O(l^a * b^r)$, where l is the maximum length of the lists in $\mathcal{M}$ (or at least 1), a is the fold rank of $\mathcal{M}$, b is the maximum size of the declaration and rule bodies in $\mathcal{M}$, and r is the definition rank of $\mathcal{M}$.*

Example 4. The fold rank of the n-queens model presented in example 1 is 2. Therefore the size of the generated program is in $O(l^2)$. The bound is tight in this example.

Example 5. The exponential size of the generated code in the definition rank of the model can be reached with the following model:

```
rule c1(A) := c2(A+1) and c2(2*A)
rule c2(A) := c3(A+1) and c3(2*A)
...
rule cn(A) := c(A+1) and c(2*A)
rule c(A) := A # 666
```

In this example, the generated code for the query $c1(X)$ is of size 2^n.

4   The Dynamic Compilation Schema

The dynamic compilation schema is defined by two transformations which produce intermediate code. The first transformation, noted $-\!\langle\text{dyn}\rangle\!\longrightarrow$, expands a query to a deterministic code which adds the constraints and calls the dynamic search part. The second transformation, noted $-\!\langle{}^{\text{dyn}}_{\text{srch}}\rangle\!\longrightarrow$, rewrites the search part to a non-deterministic code which performs the reordering and search. The intermediate code follows the syntax of Rules2CP programs but allows recursion. Search-tree directives $\mathcal{S}$ are eliminated and reformulated by $-\!\langle{}^{\text{dyn}}_{\text{srch}}\rangle\!\longrightarrow$.

It is worth noting that the operator **or** represents a reified $\vee$-constraint in the deterministic code, and a choice-point in the non-deterministic code. The syntactic construction `delay(p(X))` is introduced in the intermediate code to denote the symbolic term $p(X)$ as opposed to a call to the definition $p(X)$. Such an intermediate code is then straightforward to translate to a Prolog or Java program.

To illustrate dynamic compilation, let us consider two rule definitions that constrain the shape of objects in a simple two-dimensional placement problem of thin sticks, where the sticks can be either short (from 1 to 5 units), normal (from 11 to 15 units) or long (from 21 to 25 units). A stick is a 1-unit wide rectangle which can be either horizontal or vertical.

```
shape_constraint(O) = exists(S, [1, 11 , 21],
                        shape_stick(O, S, S + 4)).
shape_stick(O, Min, Max) = domain(O:w, Min, Max) and O:h = 1
                      or domain(O:h, Min, Max) and O:w = 1.
```

The compilation scheme for `fold` described in the next section transforms the expression `shape_constraint(S)` into a code computing the same answers as the following unfolded expression:

$$((1{\leq}S{:}w \text{ and } S{:}w{\leq}1{+}4) \text{ and } S{:}h{=}1) \text{ or } ((1{\leq}S{:}h \text{ and } S{:}h{\leq}1{+}4) \text{ and } S{:}w{=}1)$$
$$\text{or } (((11{\leq}S{:}w \text{ and } S{:}w{\leq}11{+}4) \text{ and } S{:}h{=}1) \text{ or } ((11{\leq}S{:}h \text{ and } S{:}h{\leq}11{+}4) \text{ and } S{:}w{=}1)$$
$$\text{or } (((21{\leq}S{:}w \text{ and } S{:}w{\leq}21{+}4) \text{ and } S{:}h{=}1) \text{ or } ((21{\leq}S{:}h \text{ and } S{:}h{\leq}21{+}4) \text{ and } S{:}w{=}1) \tag{1}$$
$$\text{or false})).$$

4.1   Transformation of the Query to Deterministic Code

$V \vdash \cdot -\!\langle\text{dyn}\rangle\!\longrightarrow \cdot$ reformulates search directives inductively over the structure of Rules2CP expressions as follows. V is supposed to contain all the free variables appearing in the expression: V is used to pass the context to auxiliary definitions introduced by the translation.

Each definition $p(X) = e$ is translated in the intermediate code to the definition: $p_d(X) = e'$, where $\text{fv}(e) \vdash e -\!\langle\text{dyn}\rangle\!\longrightarrow e'$. Then, translated calls rely on these definitions: $V \vdash p(X) -\!\langle\text{dyn}\rangle\!\longrightarrow p_d(X)$

Recursive predicates iterating on lists are generated for each `fold`.

$$\frac{V \vdash l -\!\langle\text{dyn}\rangle\!\longrightarrow l' \quad V \vdash i -\!\langle\text{dyn}\rangle\!\longrightarrow i'}{V \vdash \text{foldl}(A \text{ from } i, \ X \text{ in } l, \ e) -\!\langle\text{dyn}\rangle\!\longrightarrow q(l', \ i', \ V)}$$

with q a new predicate symbol described by the following definitions, where all variables are fresh with respect to V:

$$q(\text{[]}, \ I, \ \boldsymbol{V}) = I.$$
$$q(\text{[}H \mid T\text{]}, \ I, \ \boldsymbol{V}) = q(T, \ e', \ \boldsymbol{V}). \quad \boldsymbol{V} \vdash e[A := I, X := H] \longrightarrow\!\langle \mathrm{dyn}\rangle\!\longrightarrow e'$$

Other cases for $\longrightarrow\!\langle \mathrm{dyn}\rangle\!\longrightarrow$ are defined homomorphically with respect to sub-expressions, taking care of scopes and name clashes: e.g.,

$$\frac{\boldsymbol{V} \vdash d \longrightarrow\!\langle \mathrm{dyn}\rangle\!\longrightarrow d' \qquad \boldsymbol{V} \cdot X \vdash e[V := X] \longrightarrow\!\langle \mathrm{dyn}\rangle\!\longrightarrow e'}{\boldsymbol{V} \vdash \texttt{let}(V \ = \ d \ \texttt{in} \ e) \longrightarrow\!\langle \mathrm{dyn}\rangle\!\longrightarrow \texttt{let}(X \ = \ d' \ \texttt{in} \ e')}$$

where X is a fresh variable.

Search directives rely on the search transformation (defined in Sec. 4.2).

$$\frac{\boldsymbol{V} \vdash h \longrightarrow\!\langle \mathrm{dyn}\rangle\!\longrightarrow \texttt{conjunctive}(o_\wedge^1) \ \dots \ \texttt{and} \ \texttt{conjunctive}(o_\wedge^n) \ \texttt{and} \atop \texttt{disjunctive}(o_\vee^1) \ \dots \ \texttt{and} \ \texttt{disjunctive}(o_\vee^m) \atop ([o_\wedge^1, \dots, o_\wedge^n], [o_\vee^1, \dots, o_\vee^m]); \boldsymbol{V} \vdash e \longrightarrow\!\langle {\mathrm{dyn} \atop \mathrm{srch}}\rangle\!\longrightarrow e'}{\boldsymbol{V} \vdash \texttt{search}(h, \ e) \longrightarrow\!\langle \mathrm{dyn}\rangle\!\longrightarrow e'}$$

4.2   Transformation of the Search to Non-deterministic Code

The compilation of the search-strategy relies on the notion of *O-layers* in a tree: for $O \in \{\wedge, \vee\}$, we call *O-layer* of an $\wedge/\vee$-tree any maximal tree sub-graph with either only $\wedge$-nodes or only $\vee$-nodes.

The following $\wedge/\vee$-tree corresponds to the expression (1) given in the previous section, where layers have been circled:

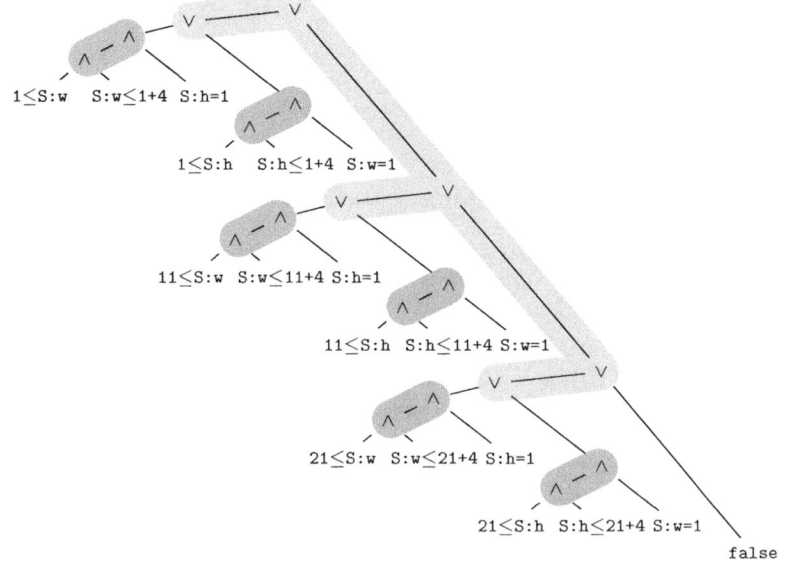

The definition of O-layers is generalized for Rules2CP expression syntax trees, by letting layers go through let-bindings, definition calls, and in the right-hand side of `implies` and through the tree intentionnaly constructed by `fold`. The child nodes of a layer are the nodes which are child of a node in the layer without being themselve in the layer. The *root O-layer* is the O-layer containing the root node if it is not the dual of O, or the empty layer otherwise. By convention, the root node is the (only) child of the empty layer. Tree reordering is applied between all the child nodes of each O-layer: criteria defined for $O \in \{\land, \lor\}$ associate a vector of scores to each child and children are reordered according to their scores, lexicographically (the score returned by the first criterion for O is considered first, then, in case of equality, the score of the second criterion for O, and so on).

Neither the tree (due to `fold` over arbitrary lists) nor the scores (due to indexicals) are supposed to be completely known at compile-time. Therefore, the transformation generates code for computing the reordering at execution-time rather than computing the reordering statically.

For a fixed pair of criteria $(o_\land, o_\lor)$, $(o_\land, o_\lor); V \vdash \cdot -\!\langle {}^{\mathrm{dyn}}_{\mathrm{srch}} \rangle\!\!\longrightarrow \cdot$ produces code which reorders the root O-layer of the tree and explores its children sequentially. $c_\land$ and $c_\lor$ are current score vectors (they have the same dimension than $o_\land$ and $o_\lor$ respectively). Initially, scores are $c_\land^{-\infty}$ and $c_\lor^{-\infty}$, vectors whose every component equals to `bottom`, since no criteria apply outside any definition. $V \vdash \cdot -\!\langle {}^{\mathrm{dyn}}_{\mathrm{srch}} \rangle\!\!\longrightarrow \cdot$ is arbitrarily defined as $(c_\land^{-\infty}, c_\lor^{-\infty}); V \vdash \cdot -\!\langle {}^{\mathrm{dyn}}_{\mathrm{srch}(\land)} \rangle\!\!\longrightarrow \cdot$ to initiate the transformation (the root layer, possibly empty, can always be considered as being an $\land$-layer). $\cdot -\!\langle {}^{\mathrm{dyn}}_{\mathrm{srch}(O)} \rangle\!\!\longrightarrow \cdot$ relies on the auxiliary transformation $(c_\land, c_\lor); V \vdash \cdot -\!\langle {}^{\mathrm{dyn}}_{\mathrm{list}(O)} \rangle\!\!\longrightarrow \cdot$ which produces code computing an associative list: for each child node of the O-layer, the score vector of the node is associated to the definition to call to explore the child recursively.

$$\frac{(c_\land, c_\lor); V \vdash e -\!\langle {}^{\mathrm{dyn}}_{\mathrm{list}(O)} \rangle\!\!\longrightarrow e'}{(c_\land, c_\lor); V \vdash e -\!\langle {}^{\mathrm{dyn}}_{\mathrm{srch}(O)} \rangle\!\!\longrightarrow \mathtt{iter\_predicates}_O(e')}$$

where $\mathtt{iter\_predicates}_O(L)$ is an internal function which iteratively selects the item of L which has the best score, executes the associated definition, then consider the other items recursively, either in conjunction or in disjunction, according to O.

Definitions and calls. For each definition $p(X) = e$, the compilation produces two definitions in the intermediate code, one for each kind of layer:

$$\frac{(u(C_\land, o_\land, p(X)), C_\lor); \mathrm{fv}(e) \vdash e -\!\langle {}^{\mathrm{dyn}}_{\mathrm{srch}(\land)} \rangle\!\!\longrightarrow e'}{(o_\land, o_\lor); V \vdash \mathtt{rule}\ p(X)\ :=\ e -\!\langle {}^{\mathrm{dyn}}_{\mathrm{srch}(\land)} \rangle\!\!\longrightarrow p_\land(C_\land, C_\lor, X)\ =\ e'}$$

$$\frac{(C_\land, u(C_\lor, o_\lor, p(X))); \mathrm{fv}(e) \vdash e -\!\langle {}^{\mathrm{dyn}}_{\mathrm{srch}(\lor)} \rangle\!\!\longrightarrow e'}{(o_\land, o_\lor); V \vdash \mathtt{rule}\ p(X)\ :=\ e -\!\langle {}^{\mathrm{dyn}}_{\mathrm{srch}(\lor)} \rangle\!\!\longrightarrow p_\lor(C_\land, C_\lor, X)\ =\ e'}$$

where the function $u(\boldsymbol{c}, \boldsymbol{o}, p(\boldsymbol{X}))$ calculates the score vector $\boldsymbol{c}'$, where components corresponding to criteria matching $p(\boldsymbol{X})$ are updated:

$$u(\overrightarrow{\boldsymbol{c}_i}, \overrightarrow{e_i \text{ for } p_i(\boldsymbol{X}_i)}, p(\boldsymbol{X})) = \overrightarrow{\boldsymbol{c}_i'}$$

where:

$$c_i' = \begin{cases} \sigma(e_i) & \text{if } \sigma(p_i(\boldsymbol{X}_i)) = p(\boldsymbol{X}) \\ c_i & \text{otherwise} \end{cases}$$

Calls rely on one of these two definitions, depending on the kind of the current layer.

$$(\boldsymbol{c}_\wedge, \boldsymbol{c}_\vee); \boldsymbol{V} \vdash p(\boldsymbol{X}) \; \overset{\text{dyn}}{\underset{\text{list}(O)}{\longrightarrow}} \; p_O(\boldsymbol{c}_\wedge,\ \boldsymbol{c}_\vee,\ \boldsymbol{X})$$

Boolean operators. $\overset{\text{dyn}}{\underset{\text{list}(\wedge)}{\longrightarrow}}$ aggregates lists in the root $\wedge$-layer. A new predicate q is introduced for each child node of the $\wedge$-layer.

$$\frac{(\boldsymbol{c}_\wedge, \boldsymbol{c}_\vee); \boldsymbol{V} \vdash a \; \overset{\text{dyn}}{\underset{\text{list}(\wedge)}{\longrightarrow}} \; a' \qquad (\boldsymbol{c}_\wedge, \boldsymbol{c}_\vee); \boldsymbol{V} \vdash b \; \overset{\text{dyn}}{\underset{\text{list}(\wedge)}{\longrightarrow}} \; b'}{(\boldsymbol{c}_\wedge, \boldsymbol{c}_\vee); \boldsymbol{V} \vdash a \text{ and } b \; \overset{\text{dyn}}{\underset{\text{list}(\wedge)}{\longrightarrow}} \; \texttt{append}(a',\ b')}$$

$$\frac{}{(\boldsymbol{c}_\wedge, \boldsymbol{c}_\vee); \boldsymbol{V} \vdash a \text{ or } b \; \overset{\text{dyn}}{\underset{\text{list}(\wedge)}{\longrightarrow}} \; [\{ \texttt{costs = } c_\wedge, \; \texttt{predicate = delay}(q(c_\wedge,\ c_\vee,\ \boldsymbol{V})) \; \}]}$$

where q applies the transformation recursively to the sub-$\vee$-layer (all variables are fresh with respect to $\boldsymbol{V}$):

$$q(\boldsymbol{C}_\wedge,\ \boldsymbol{C}_\vee,\ \boldsymbol{V}) = e. \qquad (\boldsymbol{C}_\wedge, \boldsymbol{C}_\vee); \boldsymbol{V} \vdash a \text{ or } b \; \overset{\text{dyn}}{\underset{\text{srch}(\vee)}{\longrightarrow}} \; e$$

Dual definitions hold for $\overset{\text{dyn}}{\underset{\text{list}(\vee)}{\longrightarrow}}$

Filtering

$$\frac{\boldsymbol{V} \vdash a \; \overset{\langle \text{dyn} \rangle}{\longrightarrow} \; a' \qquad (\boldsymbol{c}_\wedge, \boldsymbol{c}_\vee); \boldsymbol{V} \vdash b \; \overset{\text{dyn}}{\underset{\text{list}(O)}{\longrightarrow}} \; b'}{(\boldsymbol{c}_\wedge, \boldsymbol{c}_\vee); \boldsymbol{V} \vdash a \text{ implies } b \; \overset{\text{dyn}}{\underset{\text{list}(O)}{\longrightarrow}} \; \texttt{filter}(c_O,\ a',\ b')}$$

where, $\texttt{filter}(\boldsymbol{c},\ e,\ e')$ is an internal function which returns e' if e is true, and returns the singleton list $[\{ \texttt{costs = } \boldsymbol{c}, \texttt{ predicate = delay(true) } \}]$ otherwise.

Let-binding

$$\frac{\boldsymbol{V} \vdash v \xrightarrow{\langle \text{dyn} \rangle} v' \qquad (\boldsymbol{c}_\wedge, \boldsymbol{c}_\vee); \boldsymbol{V} \cdot Y \vdash e[X := Y] \xrightarrow{\langle \substack{\text{dyn} \\ \text{list}(O)} \rangle} e'}{(\boldsymbol{c}_\wedge, \boldsymbol{c}_\vee); \boldsymbol{V} \vdash \texttt{let}(X \ = \ v \ \texttt{in} \ e) \xrightarrow{\langle \substack{\text{dyn} \\ \text{list}(O)} \rangle} \texttt{let}(Y \ = \ v', \ e')}$$

where Y is a fresh variable.

Aggregators. Aggregators use a special source symbol, rec, to handle recursion.

$$\frac{\boldsymbol{V} \vdash \texttt{reverse}(l) \xrightarrow{\langle \text{dyn} \rangle} l'}{(\boldsymbol{c}_\wedge, \boldsymbol{c}_\vee); \boldsymbol{V} \vdash \texttt{foldl}(A \ \texttt{from} \ i, \ X \ \texttt{in} \ l, \ e) \xrightarrow{\langle \substack{\text{dyn} \\ \text{list}(O)} \rangle} q_O(l', \ \boldsymbol{c}_\wedge, \ \boldsymbol{c}_\vee, \ \boldsymbol{V})}$$

where q_O is a new predicate symbol described by the following definitions (all variables are fresh with respect to $\boldsymbol{V}$):

$$q_O(\texttt{[]}, C_\wedge, C_\vee, V) = i'. \qquad \qquad \frac{(C_\wedge, C_\vee); \boldsymbol{V} \vdash i \xrightarrow{\langle \substack{\text{dyn} \\ \text{list}(O)} \rangle} i'}{}$$

$$q_O(\texttt{[}H \ | \ T\texttt{]}, C_\wedge, C_\vee, V) = e'. \qquad \frac{(C_\wedge, C_\vee); \boldsymbol{V} \cdot H \vdash}{e[A := \texttt{rec}(q, \ T, \ \boldsymbol{V}), X := H]} \xrightarrow{\langle \substack{\text{dyn} \\ \text{list}(O)} \rangle} e'$$

and rec is translated to a recursive call to q:

$$(\boldsymbol{c}_\wedge, \boldsymbol{c}_\vee); \boldsymbol{V} \cdot H \vdash \texttt{rec}(q, \ T, \ \boldsymbol{V}) \xrightarrow{\langle \substack{\text{dyn} \\ \text{list}(O)} \rangle} q_O(T, \ \boldsymbol{c}_\wedge, \ \boldsymbol{c}_\vee, \ \boldsymbol{V})$$

Constraints and sub-search directives. Constraints and sub-search directives are children of the layer, therefore the transformation produces singleton lists associating their score to a fresh predicate q.

$$(\boldsymbol{c}_\wedge, \boldsymbol{c}_\vee); \boldsymbol{V} \vdash e \xrightarrow{\langle \substack{\text{dyn} \\ \text{list}(O)} \rangle} \texttt{[\{ costs = } c_O, \\ \texttt{predicate = delay}(q(\boldsymbol{V})) \ \texttt{\}]}$$

where q applies the transformation recursively (all variables are fresh with respect to $\boldsymbol{V}$):

$$q(\boldsymbol{V}) = e'. \qquad \boldsymbol{V} \vdash e \xrightarrow{\langle \text{dyn} \rangle} e'$$

Property 1. There are $\mathbf{O}(d \cdot s)$ p_d-, $p_\vee$- and $p_\wedge$-definitions in intermediate code, where d is the number of definitions in the Rules2CP code and s is the number of search clauses. Each definition in the intermediate code, including the auxiliary definitions for fold and sub-layers, has a size linear in the size of the original Rules2CP definition. In particular, if there is one search clause, the intermediate code has a size linear in the size of the original Rules2CP code. The complexity of the transformation is linear in the size of the generated code.

Proof. $\xrightarrow{\langle \text{dyn} \rangle}$ and $\xrightarrow{\langle \substack{\text{dyn} \\ \text{list}} \rangle}$ are inductive transformations where each step linearly composes results of the sub-transformations, either in auxiliary definitions or in-place expressions. Therefore, there exists a multiplicative constant factor between the size of the generated definitions and the size of the original Rules2CP definition. For each Rules2CP definition $p(X)$, there is one definition p_d in the intermediate code, plus two definitions $p_\vee$ and $p_\wedge$ by search clauses.

This complexity result contrasts with Rules2CP transformation complexity[4] where definition unfolding leads to exponential code size in the worst case.

Example 6. Consider the result of transforming the 4-queens Rules2CP model by the dynamic compilation schema. Instead of expanding rule definitions as in the static schema, the dynamic schema generates one definition of the intermediate code for each definition, *e.g.* safe/1 and queens_constraints/1, as follows:

```
safe(L) =
  all_different(rcp_variables(L)) and safe_foldl_2(L, 1, []).

queens_constraints(B, N) =
  domain(rcp_variables(B), 1, N) and safe(B).
```

Similarly, one (recursive) definition is generated for each aggregator of the model, *e.g.* the two nested universal quantifiers:

```
safe_foldl_2([], I_safe_foldl_2, _) = I_safe_foldl_2.
safe_foldl_2([Q_2 | Tail_2], I_safe_foldl_2, []) =
  safe_foldl_2(Tail_2, I_safe_foldl_2 and
  safe_foldl_3(L,1,Q_2), []).

safe_foldl_3([], I_safe_foldl_3, _) = I_safe_foldl_3.
safe_foldl_3([R_3 | Tail_3], I_safe_foldl_3, [Q_2]) =
  safe_foldl_3(Tail_3,
              (I_safe_foldl_3 and
              let(I := rcp_att(Q_2, column),
                  J := rcp_att(R_3, column),
              I < J implies
              rcp_att(Q_2, row) # J - I + rcp_att(R_3, row) and
              rcp_att(Q_2, row) # I - J + rcp_att(R_3, row))),
              [Q_2]).
```

As for the search component, all rules in the scope of a search predicate generate two definitions of the intermediate code, one for a use in a conjunctive context and one for the disjunctive context.

When compiled with the dynamic schema, the model presented in the example 1 can be advantageously modified by writing the rule queens_labeling_var as follows:

```
rule queens_labeling_var(Var) :=
  exists(Val in [domain_min(Var) .. domain_max(Var)],
         queens_labeling_val(Var, Val)).
```

Here, the existential quantifier ranges over the actual bounds of queen variables instead of [1 .. N] as in the static version, thus allowing the search to benefit from propagation.

Similarly, the search tree ordering heuristics can be written with a dynamic criterion as follows:

```
heuristics mo :=
  disjunctive(
    least(abs((domain_max(Var) - domain_min(Var))/2 - Val))
    for queens_labeling_val(Var, Val)).
```

5   Evaluation

In this section, we first compare the compilation times and run times of Rules2CP and Cream. The performances are measured on classical N-Queens, Bridge Scheduling, and Open-Shop Scheduling problems. Then, we report performances of Cream on the Optimal Rectangle Packing problem which illustrates the need for dynamic search strategies that cannot be compiled with the static expansion schema.

5.1   Comparison of Both Compilation Schemes

The Bridge problem consists in finding a schedule, involving 46 tasks subject to precedence, distance and resource requirement constraints, that minimizes the time to build a five-segment bridge [14] p. 209.

The Open-Shop problem consists in finding the non-preemptive schedule with minimal completion time of a set J of n jobs, consisting each of m tasks, on a set M of m machines. The processing times are given by a $m \times n$-matrix P, in which $p_{ij} \geq 0$ is the processing time of task $T_{ij} \in T$ of job J_j to be done on machine M_i. The tasks of a job can be processed in any order, but only one at a time. Similarly, a machine can process only one task at a time. Here, the j6-4 ($n = m = 6$) and j7-1 ($n = m = 7$) Open-Shop problem instances (Brucker *et al.* [1]) are considered.

Table 3 compares the compilation and execution runtimes in seconds in Cream with those obtained in Rules2CP.

In all N-Queens instances, the "first-fail variables selection heuristics" is applied. In Rules2CP, first-fail is handled by the SICStus labeling/2 built-in, whereas in Cream selection is handled by generated code (leaning on the domain_size/1 predicate in this case).

In all scheduling problem instances, the same heuristics on disjunctive formulae with static criterion "schedule first the task that has the greatest duration" was used. The implementation of the Cream compiler is a proof of concept of the transformations presented in Sec. 4, and no effort has been made yet to improve performances.

When heuristics on formulae are involved, the compilation in Cream is about twice faster than in Rules2CP because ordering is delayed to execution time and partial evaluation does not occur.

Table 3. Rules2CP and Cream programs runtimes in seconds

	Rules2CP		Cream	
	Compilation	Solving	Compilation	Solving
8-Queens	0.070	0.000	0.020	0.000
16-Queens	0.290	0.000	0.020	0.020
32-Queens	1.840	0.005	0.020	0.080
64-Queens	15.430	0.030	0.020	0.340
96-Queens	58.510	0.060	0.020	0.740
Bridge	0.360	0.150	0.200	0.370
Open-Shop j6-4	1.370	160	0.790	325
Open-Shop j7-1	2.150	1454	1.310	2327

On the one hand, Cream yields structured constraint programs including (re-cursive) clauses as a programmer would have written the model in Prolog. On the other hand, Rules2CP produces optimized flatten constraint programs by complete expansion of definitions and record projections with partial evaluation.

Solving runtimes of constraint programs generated by Cream are twice slower than those generated by Rules2CP. This overhead is explained by the following reasons: (a) in both Rules2CP and Cream, finite domain variables are global variables. But in constraint programs generated with Cream, they are handled by a backtrackable table associating names with actual variables. Whereas programs generated by Rules2CP does not need such a mechanism because of the complete expansion scheme; (b) In Rules2CP, partial evaluation at compile-time avoids the need of Prolog tests for handling logical implication as it is the case in programs generated with Cream; (c) record projections, finite domain arithmetic expressions computation, and goal calls in general are yet other sources of overhead. As we considered optimization problems, this aggregation of overheads for one call of the search goal is to multiply by the number of iterations of the branch and bound algorithm; (d) finally, priority queues could advantageously substitute for lists of pairs to enumerate children of layers.

It is worth noticing that these points are mainly implementation details and should be avoided in future work by an optimizing compiler.

5.2   Dynamic Search Strategies

Our work on rule-base modelling languages for constraint programming originates from the EU project Net-WMS[1] which aims at solving real-size non-pure bin packing problems of the automotive industry. Three-dimensional Packing problems tackled in this project involve many business constraints, in addition to pure containment and non-overlapping constraints. To solve efficiently these problems, it is mandatory to benefit as much as possible from propagation during the search. Hence the need of expressing dynamic search strategies which depend on the values or domains of variables at runtime.

---

[1] http://net-wms.ercim.org

Table 4. Optimal Rectangle Packing problem runtimes in seconds (Linux / Intel Core2 CPU, 2.83GHz)

n	Compilation	Solving	
	Cream	Cream	Reference
18	0.650	17	9
19	0.700	17	8
20	0.780	30	17
21	0.810	100	63
22	0.870	430	297
23	0.930	2700	1939
24	0.980	3900	2887
25	1.060	27020	20713

The Optimal Rectangle Packing problem, also known as Korf's benchmark [9], consists in finding the smallest rectangle containing n squares of sizes $S_i = i$ for $1 \leq i \leq n$. In [13], Simonis and O'Sullivan have provided a simple but efficient dynamic search strategy for solving this problem in SICStus Prolog, improving the best known runtimes obtained by Korf up to a factor of 300. We have transposed their model in Cream and report the performance figures obtained with the same SICStus Prolog system in table 4.

Table 4 shows that with fast compilation times, Cream generates SICStus Prolog code nearly as efficient as the hand-written SICStus Prolog program of [13] for the different instances of the problem.

6   Conclusion

Modelling languages for stating combinatorial optimization problems can be interpreted to produce executable constraint programs by fundamentally two compilation schemas: the static expansion schema and the procedural code generation schema. We have shown that the static expansion schema may generate constraint programs of exponential size in the level of nesting of definitions (which remains limited in practice), while the code generation schema generates code of linear size. In our implementation of both schemas for the rule-based modelling language Rules2CP, we have shown that the code generation schema exhibits a time overhead of approximatively a factor 2 at runtime w.r.t. the statically expanded code. Furthermore the code generation schema makes it possible to benefit from propagation during search by executing dynamical search strategies specified in Rules2CP with a good efficiency as shown on optimal rectangle packing problems. All these results thus militate in favor of the procedural code generation schema which should probably be preferred to the static expansion schema.

The declarative specification of ordering heuristics by pattern matching on rules' left-hand sides introduced in Rules2CP should also be applicable to other modelling languages that use definitions, such as Zinc [11,3] for instance.

A natural extension for future work is the specification of more complex search procedures which are currently limited in our system to depth-first backtracking and branch and bound search.

Acknowledgements. We acknowledge support from European FP6 Strep project Net-WMS, and discussions with the partners of this project. Special thanks go to Sylvain Soliman for his insights and to Sunrinderjeet Singh who developped the Rules2CP models for Open-Shop.

References

1. Brucker, P., Hurink, J., Jurisch, B., Wöstmann, B.: A branch & bound algorithm for the open-shop problem. In: GO-II Meeting: Proceedings of the Second International Colloquium on Graphs and Optimization, pp. 43–59. Elsevier Science Publishers B. V., Amsterdam (1997)
2. Carlsson, M., Beldiceanu, N., Martin, J.: A geometric constraint over k-dimensional objects and shapes subject to business rules. In: Stuckey, P.J. (ed.) CP 2008. LNCS, vol. 5202, pp. 220–234. Springer, Heidelberg (2008)
3. de la Banda, M.G., Marriott, K., Rafeh, R., Wallace, M.: The modelling language Zinc. In: Benhamou, F. (ed.) CP 2006. LNCS, vol. 4204, pp. 700–705. Springer, Heidelberg (2006)
4. Fages, F., Martin, J.: From rules to constraint programs with the Rules2CP modelling language. In: Oddi, A., Fages, F., Rossi, F. (eds.) CSCLP 2008. LNCS, vol. 5655, pp. 66–83. Springer, Heidelberg (2009)
5. Fages, F., Martin, J.: Modelling search strategies in Rules2CP. In: van Hoeve, W.-J., Hooker, J.N. (eds.) CPAIOR 2009. LNCS, vol. 5547, pp. 321–322. Springer, Heidelberg (2009)
6. Frisch, A.M., Harvey, W., Jefferson, C., Martinez-Hernandez, B., Miguel, I.: Essence: A constraint language for specifying combinatorial problems. Constraints 13, 268–306 (2008)
7. Van Hentenryck, P., Perron, L., Puget, J.-F.: Search and strategies in opl. ACM Transactions on Compututational Logic 1(2), 285–320 (2000)
8. Kanellakis, P.C., Mitchell, J.C.: Polymorphic unification and ml typing. In: Proceedings of the 16th ACM SIGPLAN-SIGACT Symposium on Principles of Programming Languages, POPL 1989, pp. 105–115. ACM, New York (1989)
9. Korf, R.E.: Optimal rectangle packing: New results. In: ICAPS, pp. 142–149 (2004)
10. Michel, L., Van Hentenryck, P.: The comet programming language and system. In: van Beek, P. (ed.) CP 2005. LNCS, vol. 3709, p. 881. Springer, Heidelberg (2005)
11. Rafeh, R., de la Banda, M.G., Marriott, K., Wallace, M.: From Zinc to design model. In: Hanus, M. (ed.) PADL 2007. LNCS, vol. 4354, pp. 215–229. Springer, Heidelberg (2006)
12. Rémy, D.: Records and variants as a natural extension of ML. In: Sixteenth Annual Symposium on Principles of Programming Languages (1989)
13. Simonis, H., O'Sullivan, B.: Using global constraints for rectangle packing. In: Proceedings of the First Workshop on Bin Packing and Placement Constraints BPPC 2008, Associated to CPAIOR 2008 (May 2008)
14. Van Hentenryck, P.: The OPL Optimization programming Language. MIT Press, Cambridge (1999)

Solving the Static Design Routing and Wavelength Assignment Problem

Helmut Simonis*

Cork Constraint Computation Centre
Department of Computer Science, University College Cork, Ireland
h.simonis@4c.ucc.ie

Abstract. In this paper we present a hybrid model for the static design variant of the routing and wavelength assignment problem in directed networks, an important benchmark problem in optical network design. Our solution uses a decomposition into a MIP model for the routing aspect, combined with a graph coloring step modelled using either MIP (Coin-OR), SAT (minisat) or finite domain constraints (ECLiPSe). We consider two possible objective functions, one minimizing the maximal number of frequencies used on any of the links, the other minimizing the total number of frequencies used. We compare the models on a set of benchmark tests, results show that the constraint model is much more scalable than the alternatives considered, and is the only one producing proven optimal or near optimal results when minimizing the total number of wavelengths.

1 Introduction

The routing and wavelength assignment problem (RWA) [10,1,17] in optical networks considers a network where demands can be transported on different optical wavelengths through the network. Each accepted demand is allocated a path from its source to its sink, as well as a specific wavelength. Demands routed over the same link must be allocated to different wavelengths, while demands whose paths are link disjoint may use the same wavelength.

The RWA problem is a well studied, important problem in optical network design, for which many problem variants have been considered. Depending on the technology used, the network may be assumed to be *directed* or *undirected*. The *static design problem* considers the problem of allocating all given demands on the network topology, using the minimal number of frequencies. The *demand acceptance problem* considers a fixed, given number of frequencies on all links in the network. The objective is to accept the maximal number of demands in the network. In this paper we discuss the static design problem in a directed network, while a constraint-based solution for the demand acceptance problem has been described in [13].

More formally, we are considering a directed network $G = (N, E)$ of nodes N and edges E. A demand $d \in D$ is between source $s(d)$ and sink $t(d)$. We use the notation

---

* This work was supported by Science Foundation Ireland (Grant Number 05/IN/I886). Support from Cisco Systems and the Silicon Valley Community Foundation is gratefully acknowledged.

J. Larrosa and B. O'Sullivan (Eds.): CSCLP 2009, LNAI 6384, pp. 59–75, 2011.

In(n) and Out(n) to denote all edges entering resp. leaving node n. An a priori upper bound on the number of available wavelengths is required, we use the set Λ for this purpose.

Figure 1 shows one of the example networks we will use in the evaluation, with just two demands (5-13) and (1-12). On the left, the demands are allocated to different frequencies (colours), and thus can share link 8-9, on the right they use the same frequency, and therefore must be routed on link disjoint paths.

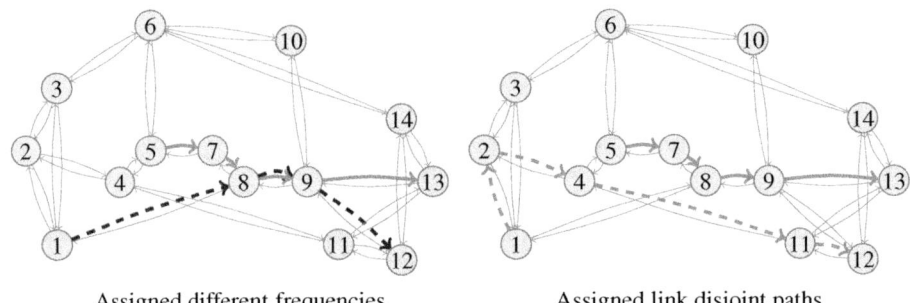

Assigned different frequencies   Assigned link disjoint paths

Fig. 1. Example Network ns f with 2 Demands

1.1   Basic Problem

We can formulate a *basic model* of the problem with two sets of 0/1 integer variables. Variables y_d^λ denote whether demand d is accepted using wavelength λ, variables x_{de}^λ state whether edge e is used to transport demand d on wavelength λ.

$$\min \max_{e \in E} \sum_{d \in D, \lambda \in \Lambda} x_{de}^\lambda \tag{1}$$

s.t.

$$y_d^\lambda \in \{0, 1\}, x_{de}^\lambda \in \{0, 1\} \tag{2}$$

$$\forall d \in D : \quad \sum_{\lambda \in \Lambda} y_d^\lambda = 1 \tag{3}$$

$$\forall e \in E, \forall \lambda \in \Lambda : \quad \sum_{d \in D} x_{de}^\lambda \le 1 \tag{4}$$

$$\forall d \in D, \forall \lambda \in \Lambda : \quad \sum_{e \in \text{In}(s(d))} x_{de}^\lambda = 0, \quad \sum_{e \in \text{Out}(s(d))} x_{de}^\lambda = y_d^\lambda \tag{5}$$

$$\forall d \in D, \forall \lambda \in \Lambda : \quad \sum_{e \in \text{Out}(t(d))} x_{de}^\lambda = 0, \quad \sum_{e \in \text{In}(t(d))} x_{de}^\lambda = y_d^\lambda \tag{6}$$

$$\forall d \in D, \forall \lambda \in \Lambda, \forall n \in N \setminus \{s(d), t(d)\} : \quad \sum_{e \in \text{In}(n)} x_{de}^\lambda = \sum_{e \in \text{Out}(n)} x_{de}^\lambda \tag{7}$$

Constraint (2) enforces integrality of the solution, constraint (3) states that all demands must be accepted and must use exactly one wavelength. The *clash* constraint (4) states that on each edge, only one demand may use any given wavelength. We further have constraints (5) and (6), which link the x and y variables at the source (resp. sink) of each demand. Finally, constraint (7) enforces flow balance on all other nodes of the network.

1.2   Extended Problem

Note that this model minimizes the maximal number of frequencies used on any link, not the overall number of frequencies. For this we have to introduce another set of 0/1 indicator variables z^λ which state whether wavelength λ is used by any demand in the network. The objective is then to minimize the sum of the z^λ variables. We also impose inequality constraints between x^λ_{de} and z^λ variables in constraint (11) of the following, *extended model* which force the indicator variable for a frequency to be set as soon as one demand uses the frequency.

It is not clear a priori whether the basic or the extended model capture the objective of minimizing the number of frequencies used, we will have to consider both alternatives in our solution approach. Both variants occur in the literature [5], without a clear indication which would be more relevant in practice.

$$\min \sum_{\lambda \in \Lambda} z^\lambda \tag{8}$$

s.t.

$$z^\lambda \in \{0,1\}, y^\lambda_d \in \{0,1\}, x^\lambda_{de} \in \{0,1\} \tag{9}$$

$$\forall d \in D: \quad \sum_{\lambda \in \Lambda} y^\lambda_d = 1 \tag{10}$$

$$\forall d \in D, \forall e \in E, \forall \lambda \in \Lambda: \quad x^\lambda_{de} \le z^\lambda \tag{11}$$

$$\forall e \in E, \forall \lambda \in \Lambda: \quad \sum_{d \in D} x^\lambda_{de} \le 1 \tag{12}$$

$$\forall d \in D, \forall \lambda \in \Lambda: \quad \sum_{e \in \mathrm{In}(s(d))} x^\lambda_{de} = 0, \quad \sum_{e \in \mathrm{Out}(s(d))} x^\lambda_{de} = y^\lambda_d \tag{13}$$

$$\forall d \in D, \forall \lambda \in \Lambda: \quad \sum_{e \in \mathrm{Out}(t(d))} x^\lambda_{de} = 0, \quad \sum_{e \in \mathrm{In}(t(d))} x^\lambda_{de} = y^\lambda_d \tag{14}$$

$$\forall d \in D, \forall \lambda \in \Lambda, \forall n \in N \setminus \{s(d), t(d)\}: \quad \sum_{e \in \mathrm{In}(n)} x^\lambda_{de} = \sum_{e \in \mathrm{Out}(n)} x^\lambda_{de} \tag{15}$$

Figure 2 shows the difference between the basic and extended cost on a small example with three nodes 1, 2, 3 and three demands A, B, C. On each directed link we need only two colours, that means that the basic model has cost 2, but overall we need three colours for a feasible solution, the extended model therefore has cost 3.

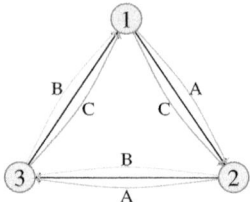

Fig. 2. Difference Between Basic and Extended Cost

1.3   Contribution and Related Work

As the complete model is quite hard to solve, it has been suggested before [1] that a two-step decomposition into a routing and a wavelength assignment phase would be a good solution technique for this problem. We re-use this idea, but strengthen it by improving each phase with some new techniques.

The main contributions of this paper are

- a comparison of different, generic solution methods for the generated graph coloring problem, using MIP, SAT and finite domain constraint programming,
- a new, very accurate lower bound to the RWA problem based on a resource-based relaxation of an existing, source aggregation MIP solution,
- experimental results showing that using constraint programming very high quality solutions are obtained by this method in seconds, significantly outperforming the other techniques,
- results indicate that the basic problem is relatively easy to solve with a variety of techniques, while the extended problem is much harder.

The RWA problem has been studied using many different solution methods, see [4] for an overview. We can distinguish two main approaches. Greedy heuristics use local search techniques to accept demands incrementally, providing fast solutions for large problem cases, but without a formal guarantee of solution quality. Alternatively, complete methods, mainly based on ILP (Integer Linear Programming) techniques, can provide optimal solutions, but are restricted in the problem size handled [5,6].

The static design problem considered here requires a somewhat different solution approach than the demand acceptance problem discussed in [13]. It uses a similar two-phase decomposition, but the relaxation of the second phase is much simpler, handled by adding additional frequencies rather than using explanation techniques to identify demands to be removed from the problem. At the same time, the resource MIP problems seems more difficult to solve for the static design case, restricting scalability with regards to network size.

A general overview of constraint applications in the network domain is given in [12]. Smith in [14] discusses a design problem for optical networks, but this is restricted to a ring topology, and minimizes the need for ADM multiplexers.

The RWA problem considered here is not too far removed from the static design problem in MPLS traffic engineering (MPLS-TE) in IP networks, which has been approached with multiple hybrid constraint solution techniques as described in [8,7,12].

The main difference is that demands in the MPLS-TE problem have integer sizes and overall link capacity limits are enforced instead of clash constraints. Note that the choice of objective function (static design vs. demand acceptance) also plays a major role in influencing the solution methods for MPLS-TE.

2   Source Aggregation

The direct formulation of the problem based on (1) or (8) does not scale well for increasing network size or number of demands. A possible improvement has been described in the literature for the RWA by aggregating flows for all demands originating in the same source node. This removes some of the symmetries that have to be considered and reduces the problem sensitivity to increasing number of demands. We can adjust the source aggregation model used in [13] based on [5] to the different objective functions discussed here, this leads to the following model for the basic problem:

$$\min z_{\max} \tag{16}$$

s.t.

$$z_{\max} \in \{0, 1 ... |\Lambda|\}, x_{se}^{\lambda} \in \{0, 1\} \tag{17}$$

$$\forall e \in E, \forall \lambda \in \Lambda: \quad \sum_{s \in N} x_{se}^{\lambda} \leq 1 \tag{18}$$

$$\forall s \in N, \forall \lambda \in \Lambda: \quad \sum_{e \in \text{In}(s)} x_{se}^{\lambda} = 0 \tag{19}$$

$$\forall s \in N, \forall d \in D_s, \forall \lambda \in \Lambda: \quad \sum_{e \in \text{In}(d)} x_{se}^{\lambda} \geq \sum_{e \in \text{Out}(d)} x_{se}^{\lambda} \tag{20}$$

$$\forall s \in N, \forall d \in D_s: \quad \sum_{\lambda \in \Lambda} \sum_{e \in \text{In}(d)} x_{se}^{\lambda} = \sum_{\lambda \in \Lambda} \sum_{e \in \text{Out}(d)} x_{se}^{\lambda} + P_{sd} \tag{21}$$

$$\forall s \in N, \forall n \neq s, n \notin D_s, \forall \lambda \in \Lambda: \quad \sum_{e \in \text{In}(n)} x_{se}^{\lambda} = \sum_{e \in \text{Out}(n)} x_{se}^{\lambda} \tag{22}$$

$$\forall e \in E: \quad \sum_{s \in N} \sum_{\lambda \in \Lambda} x_{se}^{\lambda} \leq z_{\max} \tag{23}$$

Constraints (17) define the integrality conditions. Constraint (18) specifies the *clash* constraint between demands from different sources. Constraint (19) states that demands originating in s can not be routed through s, while constraints (20) and (21) consider the destinations of demands originating in s and state that the correct number P_{sd} of demands must be dropped in each node. Constraint (22) enforces flow balance at all other nodes of the network. The integer objective value $z_{\max}$ is linked to the decision variables via the inequalities (23) which bound the cost by the maximal number of frequencies used on any link of the network.

If we change the objective function to handle the extended problem, we find that the model can no longer solve realistic problem instances.

3 Solution Approach

In this section we describe our solution approach which is based on a simple decomposition strategy already proposed in [1]. A solution to the static design RWA problem must consider the following three activities:

1. Choose path for each demand
2. Assign wavelength for each demand
3. Minimize number of wavelengths used (basic or extended model)

We choose a decomposition technique which handles the first step with a MIP program which assigns paths to the demands while minimizing the maximal number of demands routed over a link. The second and third step are expressed as a graph coloring problem where the nodes are demands and disequality constraints (edges) are imposed between any two demands which are routed over the same link in the network. The overall solution approach is shown in Figure 3. When using the MIP-MIP decomposition, the graph coloring problem is solved as an optimization problem, minimizing the number of wavelengths used. In the MIP-SAT/FD decomposition, we use a feasibility check for the graph coloring problem. We start with the minimal number of wavelengths required by the solution of the first phase. If we find a solution, the overall problem is solved to optimality. If the problem is infeasible (or the solver times out), we increase the number of wavelengths considered until we find a good, but possibly sub-optimal solution.

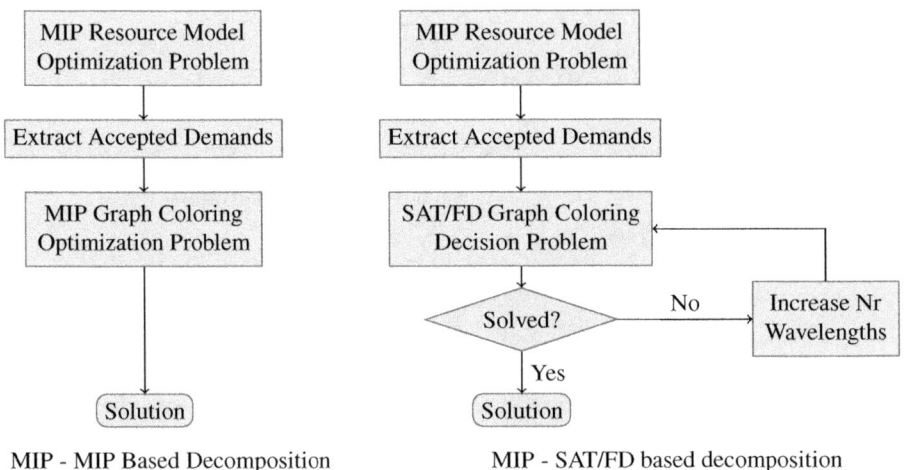

MIP - MIP Based Decomposition   MIP - SAT/FD based decomposition

Fig. 3. Solution Approach

3.1 Phase 1

The input for phase 1 is a demand matrix, an example for the nsf network is shown in Figure 4. The colours encode the minimal distance between the nodes.

Fig. 4. Sample Demand Matrix (100 Demands) for nsf Network

The first phase of the decomposition is a MIP model which minimizes the maximum number of demands routed over any link in the network. The model is a relaxation of the complete model (16), obtained by ignoring allocated frequencies and instead only counting the number of demands routed over each link. Integer variables z_{se} state how many demands originating in s are routed over edge e. The domain of these variables is limited by T_s, the total number of demands originating in s.

$$\min z_{\max} \tag{24}$$

s.t.

$$z_{\max} \in \{0, 1...|\Lambda|\}, z_{se} \in \{0, 1...T_s\} \tag{25}$$

$$\forall s \in N : \quad \sum_{e \in \text{In}(s)} z_{se} = 0 \tag{26}$$

$$\forall s \in N, \forall d \in D_s : \quad \sum_{e \in \text{In}(d)} z_{se} = \sum_{e \in \text{Out}(d)} z_{se} + P_{sd} \tag{27}$$

$$\forall s \in N, \forall n \neq s, n \notin D_s : \quad \sum_{e \in \text{In}(n)} z_{se} = \sum_{e \in \text{Out}(n)} z_{se} \tag{28}$$

$$\forall e \in E : \quad \sum_{s \in N} z_{se} \leq z_{\max} \tag{29}$$

Constraint (25) describes the integrality constraints, note that the variables have integer (not 0/1) domains. The clash constraint (4) has disappeared, the capacity limit for each link is handled as part of the objective function. Constraint (26) limits the use of the source node, while constraint (27) describes the balance around the destination nodes, using P_{sd}, the (fixed) number of demands from s to d. Finally, constraint (28) imposes flow balance for all other nodes. Constraints (29) link the objective to the decision variables.

The solution to (24) does not immediately return the routing for each demand, this requires a non-deterministic, but backtrack-free program to construct the paths, while at the same time removing possible loops from the solution. Figure 5 shows the result of phase 1 for source node 3, i.e. the third row in the demand matrix of Figure 4. Numbers in the nodes state how many demands originating in S end in that node, numbers on the edges state how many demands from the source are routed over them. The figure highlights a situation where we have multiple paths between the source S in node 3 and one of the destinations (node 11, marked A). We can freely choose which demand to send over which path, as long as we satisfy the capacity restrictions.

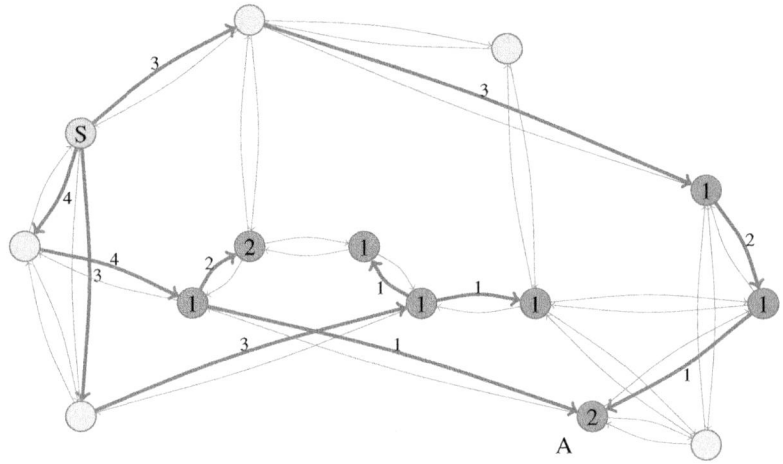

Fig. 5. Phase 1: Example Solution for Source Node 3 (Marked S)

3.2   Phase 2

The graph coloring problem for the second phase is expressed with three different solvers, a MIP optimization problem, and a SAT or finite domain decision procedure. All work on the same graph coloring instance, where each demand is a node, and two nodes (demands) are linked if they are routed over the same edge in the network. The MIP and SAT models use 0/1 integer variables x_d^λ, which state whether demand d is using wavelength λ. The finite domain model uses variables y_d which range over values 1 to Λ, the number of wavelengths considered in the model. As constraints it uses Alldifferent constraints instead of binary disequalities, which allows us to use stronger propagation methods [15].

Figure 6 shows the resource requirements computed for the sample demand matrix. The numbers (and colours) on the edges denote how many demands are routed over them, this corresponds to the size of the Alldifferent constraints required for that link. Note that the largest number of demands (13) is used on only 5 of the links. These will be the most difficult constraints to satisfy, as the number of variables is equal to the number of colours.

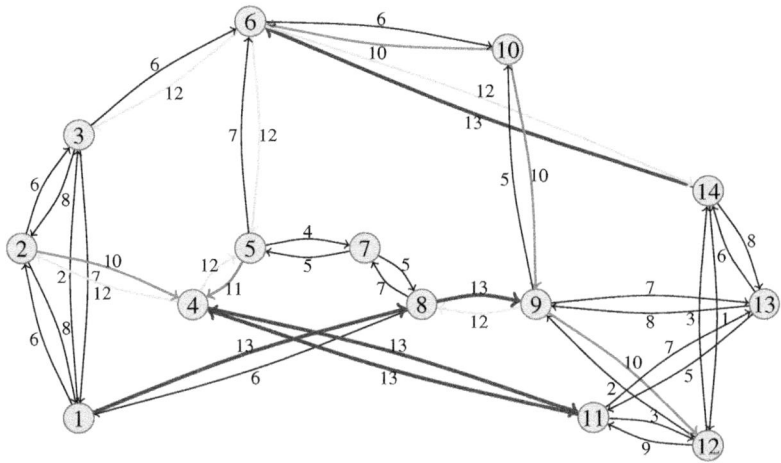

Fig. 6. Phase 2: Resource Requirements

We use the predicate $p(d, e)$ to denote if demand d was routed over edge e in the solution of the phase 1 problem.

Phase 2 MIP Formulations. The graph coloring problem leads to a simple MIP formulation for the second phase of the basic problem:

$$\min z_{\max} \tag{30}$$

s.t.

$$x_d^\lambda \in \{0, 1\}, z_{\max} \in \{0, 1, ..., |A|\} \tag{31}$$

$$\forall_{d \in D} : \sum_{\lambda \in A} x_d^\lambda = 1 \tag{32}$$

$$\forall_{e \in E} \forall_{\lambda \in A} : \sum_{\{d \in D \mid p(d,e)\}} x_d^\lambda \leq 1 \tag{33}$$

$$\forall_{e \in E} : \sum_{\lambda \in A} \sum_{\{d \in D \mid p(d,e)\}} x_d^\lambda \leq z_{\max} \tag{34}$$

The objective is to minimize the integer variable $z_{\max}$ which is bounded by the maximal number of frequencies used on any edge of the network. Constraint (32) states that each demand must be assigned to a frequency, constraint (33) imposes the clash condition that only one demand can use a given frequency on each link, and constraint (34) links the cost and the decision variables.

We can obtain a model for the extended problem by adding 0/1 decision variables z^λ which indicate if frequency λ is used by any of the demands. We link the new z^λ variables to the x_d^λ variables by inequality (40), and update the objective function with inequality (41). We keep inequalities (39), without them the linear relaxation only produces a very weak lower bound of 1.

$$\min z_{\max} \tag{35}$$

s.t.

$$x_d^\lambda \in \{0,1\}, z^\lambda \in \{0,1\}, z_{\max} \in \{0,1,...,|\Lambda|\} \tag{36}$$

$$\forall_{d\in D}: \quad \sum_{\lambda\in\Lambda} x_d^\lambda = 1 \tag{37}$$

$$\forall_{e\in E}\forall_{\lambda\in\Lambda}: \quad \sum_{\{d\in D \mid p(d,e)\}} x_d^\lambda \le 1 \tag{38}$$

$$\forall_{e\in E}: \quad \sum_{\lambda\in\Lambda}\sum_{\{d\in D \mid p(d,e)\}} x_d^\lambda \le z_{\max} \tag{39}$$

$$\forall_{d\in D}\forall_{\lambda\in\Lambda}: \quad x_d^\lambda \le z^\lambda \tag{40}$$

$$\sum_{\lambda\in\Lambda} z^\lambda \le z_{\max} \tag{41}$$

Phase 2 Finite Domain Model. For the finite domain model, we use variables y_d which range over all possible frequencies. To express the objective of the basic problem, we need to consider how many different frequencies are used on each link. We can use the NValue constraint [2] to count the number of different values used, leading to a model:

$$\min \max_{e\in E} n_e \tag{42}$$

s.t.

$$y_d \in \{0,1...,|\Lambda|\}, n_e \in \{0,1...,|\Lambda|\} \tag{43}$$

$$\forall_{e\in E}: \quad \text{nvalue}(n_e, \{y_d \mid p(d,e)\}) \tag{44}$$

$$\forall_{e\in E}: \quad \text{alldifferent}(\{y_d \mid p(d,e)\}) \tag{45}$$

Since the NValue and Alldifferent constraints are expressed over the same variable sets, the problem can be drastically simplified. We know that the values in the Alldifferent constraint must be pairwise different, and therefore find that the number of different values is equal to the number of variables in the constraint. The largest Alldifferent constraint will be set up on some link where the optimal cost was reached in phase1. The finite domain model for the basic problem therefore is no longer an optimization problem, but a feasibility problem over arbitrary domains:

$$y_d \in \{0,1...,|\Lambda|\} \tag{46}$$

$$\forall_{e\in E}: \quad \text{alldifferent}(\{y_d \mid p(d,e)\}) \tag{47}$$

For the extended problem, the phase 2 finite domain model is

$$\min \max_{d\in D} y_d \tag{48}$$

s.t.

$$y_d \in \{0,1...,|\Lambda|\} \tag{49}$$

$$\forall_{e\in E}: \quad \text{alldifferent}(\{y_d \mid p(d,e)\}) \tag{50}$$

We use "optimization from below", and try out increasing values C for the objective until we find a feasible solution. The (fixed) objective serves as upper bound on the domain of the y_d variables for each of the instances tested:

$$y_d \in \{0, 1..., C\} \tag{51}$$

$$\forall_{e \in E} : \quad \text{alldifferent}(\{y_d \mid p(d, e)\}) \tag{52}$$

Phase 2 SAT Formulation. A SAT model for the second phase can be derived using the x_d^λ variables and the clauses

$$\forall_{d \in D} \forall_{\lambda_1, \lambda_2 \in \Lambda \text{ s.t. } \lambda_1 \neq \lambda_2} : \quad \neg x_d^{\lambda_1} \vee \neg x_d^{\lambda_2} \tag{53}$$

$$\forall_{d \in D} : \quad \bigvee_{\lambda \in \Lambda} x_d^\lambda \tag{54}$$

$$\forall_{e \in E} \forall_{\lambda \in \Lambda}, d_1, d_2 \in D \text{ s.t. } p(d_1, e) \wedge p(d_2, e) \wedge d_1 \neq d_2 : \quad \neg x_{d_1}^\lambda \vee \neg x_{d_2}^\lambda \tag{55}$$

Constraints (53) state that a demand can not be assigned to more than one frequency, constraints (54) impose the other condition that each demand must be allocated to at least one frequency, and constraints (55) impose the clash constraints between any two demands routed over the same edge of the network. Alternatively, instead of the clausal representation, it is also possible to use the linear constraints of the MIP model directly in a Pseudo-Boolean solver.

4 Experimental Results

Most of the published results on the RWA problem use randomly generated demands on a few given network structures. We also use this approach and generate given numbers of demands between randomly chosen source and sink nodes. Multiple demands between the same nodes are allowed, but source and sink must be different.

In the literature we found four actual optical network topologies used in experiments. Their size is quite small, ranging from 14 to 27 nodes.

nsf 14 nodes, 42 edges
eon 20 nodes, 78 edges
mci 19 nodes, 64 edges
brezil 27 nodes, 140 edges

We explored different combinations of number of demands (100-600 demands in increments of 100) for each of the network topologies and created 100 random problem instances for each combination.

4.1 Basic Problem

The basic problem seems to be relatively well behaved, Table 1 shows some results for the complete (non-decomposed), source aggregation MIP model described in section 2.

Table 1. Selected Full MIP Examples (Basic Problem, 100 Runs Each)

Network	Dem.	λ	Opt.	Avg LP	Avg MIP	Avg LP Gap	Max LP Gap	Avg LP Time	Max LP Time	Avg MIP Time	Max MIP Time
brezil	100	50	100	4.24	4.57	0.33	0.90	165.65	686.55	277.14	1139.03
brezil	200	50	15	7.62	7.93	0.32	0.75	585.18	2022.48	861.74	2301.67
eon	100	50	100	6.36	6.65	0.29	0.75	13.69	43.94	33.62	70.92
eon	200	50	100	11.54	11.77	0.23	0.75	27.17	147.25	65.51	257.97
eon	300	50	100	16.62	16.89	0.27	0.75	33.08	143.49	121.27	517.50
eon	400	50	100	21.47	21.85	0.38	0.75	19.87	92.49	116.64	363.53
eon	500	50	100	26.43	26.62	0.19	0.75	23.44	99.09	162.55	468.56
eon	600	50	100	31.36	31.63	0.27	0.75	28.94	73.19	232.91	542.83
mci	100	50	100	7.67	7.81	0.14	0.83	8.45	26.36	20.27	42.42
mci	200	50	100	13.42	13.58	0.16	0.80	13.45	45.88	38.79	161.02
mci	300	50	100	19.24	19.37	0.13	0.80	15.56	97.48	55.78	239.11
mci	400	50	100	25.00	25.14	0.14	0.80	18.37	58.34	109.85	484.69
mci	500	50	100	30.45	30.58	0.13	0.80	16.46	50.08	129.90	454.33
mci	600	50	100	36.00	36.11	0.11	0.80	27.50	99.06	257.70	599.44
nsf	100	50	100	7.97	8.38	0.41	0.90	3.09	5.22	8.17	14.55
nsf	200	50	100	15.06	15.45	0.39	0.75	3.36	5.44	12.75	29.45
nsf	300	50	100	21.96	22.29	0.33	0.75	3.20	5.45	17.01	39.17
nsf	400	50	100	28.81	29.18	0.37	0.75	3.49	6.31	27.36	78.66
nsf	500	50	100	35.79	36.13	0.34	0.79	4.97	11.75	54.60	125.30
nsf	600	50	100	42.52	42.94	0.42	0.75	8.06	15.84	88.72	272.26

Table 2. Selected MIP-MIP Decomposition Examples (Basic Problem, 100 Runs Each)

Network	Dem.	λ	Opt.	Avg LP	Avg MIP	Avg MIP2	Max LP Gap	Max MIP2 Gap	Avg MIP Time	Max MIP Time	Avg MIP2 Time	Max MIP2 Time
brezil	100	150	100	4.24	4.57	4.57	0.90	0.00	0.41	0.59	0.91	1.53
brezil	200	150	100	7.92	8.26	8.26	0.75	0.00	0.46	0.58	4.45	5.97
brezil	300	150	100	11.51	11.92	11.92	0.80	0.00	0.47	0.63	8.08	9.64
brezil	400	150	100	15.10	15.45	15.45	0.75	0.00	0.51	0.70	10.93	15.84
brezil	500	150	100	18.76	19.10	19.10	0.75	0.00	0.48	0.64	13.09	17.84
brezil	600	150	100	22.32	22.61	22.61	0.75	0.00	0.51	0.66	16.77	20.56
eon	100	150	100	6.36	6.65	6.65	0.75	0.00	0.13	0.16	1.51	3.03
eon	200	150	100	11.54	11.77	11.77	0.75	0.00	0.14	0.19	5.27	7.66
eon	300	150	100	16.62	16.89	16.89	0.75	0.00	0.14	0.19	5.60	8.56
eon	400	150	100	21.47	21.85	21.85	0.75	0.00	0.17	0.22	7.38	12.11
eon	500	150	100	26.43	26.62	26.62	0.75	0.00	0.15	0.17	9.58	17.89
eon	600	150	99	31.36	31.63	31.63	0.75	0.00	0.17	0.20	14.04	27.50
mci	100	150	100	7.67	7.81	7.81	0.83	0.00	0.08	0.26	2.08	3.13
mci	200	150	100	13.42	13.58	13.58	0.80	0.00	0.09	0.09	5.36	7.69
mci	300	150	100	19.24	19.37	19.37	0.80	0.00	0.09	0.11	5.83	7.73
mci	400	150	100	25.00	25.14	25.14	0.80	0.00	0.10	0.13	8.71	12.76
mci	500	150	100	30.45	30.58	30.58	0.80	0.00	0.10	0.13	13.89	22.41
mci	600	150	100	36.00	36.11	36.11	0.80	0.00	0.11	0.14	22.56	43.58
nsf	100	150	100	7.97	8.38	8.38	0.90	0.00	0.04	0.05	2.38	3.64
nsf	200	150	100	15.06	15.45	15.45	0.75	0.00	0.04	0.06	1.81	4.39
nsf	300	150	100	21.96	22.29	22.29	0.75	0.00	0.04	0.06	1.98	6.33
nsf	400	150	100	28.81	29.18	29.18	0.75	0.00	0.06	0.08	3.54	12.34
nsf	500	150	100	35.79	36.13	36.13	0.79	0.00	0.05	0.06	5.77	9.38
nsf	600	150	100	42.52	42.94	42.94	0.75	0.00	0.06	0.08	9.09	16.19

Multiple problem instances for the brezil network with 200 or more demands did not find feasible solutions within 1 hour, for the other example networks solutions were found within 10 minutes.

The decomposition seems to work very well for the basic problem. Table 2 shows results for the MIP-MIP decomposition, Table 3 shows results for the MIP-FD decomposition, which finds the optimal solution for nearly all instances in less than a second. The SAT model (results shown in Table 4) is nearly as efficient.

Table 3. Selected Finite Domain Examples (Basic Problem, 100 Runs Each)

Network	Dem.	λ	Opt.	Avg LP	Avg MIP	Avg FD	Max LP Gap	Max FD Gap	Avg MIP Time	Max MIP Time	Avg FD Time	Max FD Time
brezil	100	150	100	4.24	4.57	4.57	0.90	0.00	0.43	0.59	0.01	0.02
brezil	200	150	100	7.92	8.26	8.26	0.75	0.00	0.47	0.58	0.03	0.05
brezil	300	150	99	11.51	11.92	11.93	0.80	1.00	0.49	0.63	0.07	0.09
brezil	400	150	100	15.10	15.45	15.45	0.75	0.00	0.48	0.69	0.13	0.16
brezil	500	150	100	18.76	19.10	19.10	0.75	0.00	0.48	0.64	0.23	0.27
brezil	600	150	100	22.32	22.61	22.61	0.75	0.00	0.49	0.64	0.31	0.36
eon	100	150	100	6.36	6.65	6.65	0.75	0.00	0.15	0.17	0.01	0.02
eon	200	150	100	11.54	11.77	11.77	0.75	0.00	0.15	0.19	0.04	0.06
eon	300	150	100	16.62	16.89	16.89	0.75	0.00	0.16	0.19	0.09	0.11
eon	400	150	100	21.47	21.85	21.85	0.75	0.00	0.16	0.17	0.16	0.20
eon	500	150	100	26.43	26.62	26.62	0.75	0.00	0.16	0.17	0.29	0.33
eon	600	150	100	31.36	31.63	31.63	0.75	0.00	0.16	0.19	0.40	0.47
mci	100	150	100	7.67	7.81	7.81	0.83	0.00	0.10	0.19	0.01	0.02
mci	200	150	100	13.42	13.58	13.58	0.80	0.00	0.10	0.13	0.05	0.06
mci	300	150	100	19.24	19.37	19.37	0.80	0.00	0.10	0.13	0.10	0.13
mci	400	150	100	25.00	25.14	25.14	0.80	0.00	0.11	0.13	0.19	0.20
mci	500	150	100	30.45	30.58	30.58	0.80	0.00	0.11	0.13	0.29	0.41
mci	600	150	100	36.00	36.11	36.11	0.80	0.00	0.11	0.13	0.45	0.55
nsf	100	150	100	7.97	8.38	8.38	0.90	0.00	0.05	0.06	0.02	0.02
nsf	200	150	100	15.06	15.45	15.45	0.75	0.00	0.06	0.06	0.05	0.06
nsf	300	150	100	21.96	22.29	22.29	0.75	0.00	0.06	0.06	0.10	0.13
nsf	400	150	100	28.81	29.18	29.18	0.75	0.00	0.06	0.08	0.17	0.20
nsf	500	150	100	35.79	36.13	36.13	0.79	0.00	0.06	0.06	0.31	0.34
nsf	600	150	100	42.52	42.94	42.94	0.75	0.00	0.06	0.06	0.43	0.48

Table 4. Selected SAT Examples (Basic Problem, 100 Runs Each)

Network	Dem.	λ	Opt.	Avg LP	Avg MIP	Avg SAT	Max LP Gap	Max SAT Gap	Avg MIP Time	Max MIP Time	Avg SAT Time	Max SAT Time
brezil	100	150	100	4.24	4.57	4.57	0.90	0.00	0.37	0.59	0.03	0.05
brezil	200	150	100	7.92	8.26	8.26	0.75	0.00	0.41	0.59	0.07	0.09
brezil	300	150	100	11.51	11.92	11.92	0.80	0.00	0.41	0.59	0.15	0.20
brezil	400	150	100	15.10	15.45	15.45	0.75	0.00	0.43	0.56	0.27	0.38
brezil	500	150	100	18.76	19.10	19.10	0.75	0.00	0.42	0.52	0.44	0.58
brezil	600	150	100	22.32	22.61	22.61	0.75	0.00	0.42	0.58	0.69	0.88
eon	100	150	100	6.36	6.65	6.65	0.75	0.00	0.14	0.16	0.04	0.06
eon	200	150	100	11.54	11.77	11.77	0.75	0.00	0.14	0.17	0.10	0.13
eon	300	150	100	16.62	16.89	16.89	0.75	0.00	0.14	0.17	0.24	0.31
eon	400	150	100	21.47	21.85	21.85	0.75	0.00	0.13	0.16	0.45	0.61
eon	500	150	100	26.43	26.62	26.62	0.75	0.00	0.13	0.25	0.76	1.08
eon	600	150	100	31.36	31.63	31.63	0.75	0.00	0.14	0.31	1.20	1.73
mci	100	150	100	7.67	7.81	7.81	0.83	0.00	0.13	0.23	0.05	0.08
mci	200	150	100	13.42	13.58	13.58	0.80	0.00	0.10	0.13	0.12	0.17
mci	300	150	100	19.24	19.37	19.37	0.80	0.00	0.09	0.13	0.29	0.42
mci	400	150	100	25.00	25.14	25.14	0.80	0.00	0.10	0.13	0.56	0.78
mci	500	150	100	30.45	30.58	30.58	0.80	0.00	0.10	0.27	0.97	1.41
mci	600	150	100	36.00	36.11	36.11	0.80	0.00	0.10	0.25	1.55	2.33
nsf	100	150	100	7.97	8.38	8.38	0.90	0.00	0.06	0.11	0.05	0.08
nsf	200	150	100	15.06	15.45	15.45	0.75	0.00	0.05	0.06	0.15	0.19
nsf	300	150	100	21.96	22.29	22.29	0.75	0.00	0.05	0.06	0.35	0.44
nsf	400	150	100	28.81	29.18	29.18	0.75	0.00	0.05	0.06	0.71	0.92
nsf	500	150	100	35.79	36.13	36.13	0.79	0.00	0.05	0.17	1.26	1.55
nsf	600	150	100	42.52	42.94	42.94	0.75	0.00	0.05	0.06	2.07	2.42

4.2 Extended Problem

The problem seems to be much more difficult if we consider the extended formulation minimizing the total number of frequencies used. The complete source aggregation model is not able to solve problem instances of the given sizes consistently. Table 5 shows some results for the MIP-MIP decomposition run with a timeout of 1000 seconds. After the timeout we either use the best feasible, integer solution or declare the problem as unsolved if no feasible solution has been found.

Table 5. Selected MIP-MIP Decomposition Examples (Extended Problem, 100 Runs Each)

Network	Dem.	λ	Opt.	Avg LP	Avg MIP	Avg MIP2	Max LP Gap	Max MIP2 Gap	Avg MIP Time	Max MIP Time	Avg MIP2 Time	Max MIP2 Time
brezil	100	50	94	4.24	4.57	4.63	0.90	1.00	0.35	0.53	53.59	962.72
brezil	200	50	99	7.92	8.26	8.27	0.75	1.00	0.38	0.52	141.04	331.05
brezil	300	50	88	11.48	11.87	11.94	0.80	2.00	0.38	0.50	444.64	995.14
eon	100	50	100	6.36	6.65	6.65	0.75	0.00	0.13	0.16	19.70	61.98
eon	200	50	100	11.54	11.77	11.77	0.75	0.00	0.14	0.17	188.55	925.44
mci	100	50	100	7.67	7.81	7.81	0.83	0.00	0.09	0.11	26.27	79.55
mci	200	50	96	13.42	13.58	13.63	0.80	2.00	0.10	0.13	271.65	992.88
nsf	100	50	99	7.97	8.38	8.39	0.90	1.00	0.05	0.06	29.43	967.70
nsf	200	50	99	15.06	15.45	15.46	0.75	1.00	0.05	0.06	208.72	998.00

Table 6. Selected Finite Domain Examples (Extended Problem, 100 Runs Each)

Network	Dem.	λ	Opt.	Avg LP	Avg MIP	Avg FD	Max LP Gap	Max FD Gap	Avg MIP Time	Max MIP Time	Avg FD Time	Max FD Time
brezil	100	150	95	4.24	4.57	4.62	0.90	1.00	0.44	0.61	0.02	0.11
brezil	200	150	99	7.92	8.26	8.27	0.75	1.00	0.48	0.59	0.06	0.09
brezil	300	150	99	11.51	11.92	11.94	0.80	2.00	0.49	0.64	0.12	0.19
brezil	400	150	99	15.10	15.45	15.46	0.75	1.00	0.50	0.69	0.23	0.31
brezil	500	150	96	18.76	19.10	19.16	0.75	3.00	0.50	0.66	0.93	60.63
brezil	600	150	97	22.32	22.61	22.64	0.75	1.00	0.51	0.64	0.45	0.97
eon	100	150	100	6.36	6.65	6.65	0.75	0.00	0.15	0.16	0.02	0.03
eon	200	150	100	11.54	11.77	11.77	0.75	0.00	0.16	0.19	0.07	0.16
eon	300	150	100	16.62	16.89	16.89	0.75	0.00	0.16	0.19	0.16	0.24
eon	400	150	100	21.47	21.85	21.85	0.75	0.00	0.16	0.19	0.26	0.38
eon	500	150	100	26.43	26.62	26.62	0.75	0.00	0.17	0.22	0.44	0.64
eon	600	150	100	31.36	31.63	31.63	0.75	0.00	0.17	0.20	0.60	0.98
mci	100	150	100	7.67	7.81	7.81	0.83	0.00	0.10	0.19	0.02	0.05
mci	200	150	100	13.42	13.58	13.58	0.80	0.00	0.10	0.13	0.08	0.13
mci	300	150	100	19.24	19.37	19.37	0.80	0.00	0.11	0.13	0.17	0.55
mci	400	150	100	25.00	25.14	25.14	0.80	0.00	0.11	0.14	0.32	0.58
mci	500	150	100	30.45	30.58	30.58	0.80	0.00	0.11	0.14	0.48	1.47
mci	600	150	100	36.00	36.11	36.11	0.80	0.00	0.12	0.14	0.68	1.20
nsf	100	150	99	7.97	8.38	8.39	0.90	1.00	0.06	0.08	0.03	0.03
nsf	200	150	100	15.06	15.45	15.45	0.75	0.00	0.06	0.06	0.07	0.13
nsf	300	150	100	21.96	22.29	22.29	0.75	0.00	0.06	0.06	0.15	0.19
nsf	400	150	100	28.81	29.18	29.18	0.75	0.00	0.06	0.08	0.26	0.55
nsf	500	150	100	35.79	36.13	36.13	0.79	0.00	0.06	0.08	0.42	0.53
nsf	600	150	100	42.52	42.94	42.94	0.75	0.00	0.06	0.08	0.58	0.74

Table 6 shows the result for the finite domain solver. The entries summarize the results over 100 runs with the same parameters, but different random seeds. The column *Opt.* tells how many solutions were proven optimal. The columns *Avg LP*, *Avg MIP* and *Avg FD* show the average cost obtained by the LP relaxation of the first phase MIP

Table 7. Selected SAT Examples (Extended Problem, 100 Runs Each)

Network	Dem.	λ	Opt.	Avg LP	Avg MIP	Avg SAT	Max LP Gap	Max SAT Gap	Avg MIP Time	Max MIP Time	Avg SAT Time	Max SAT Time
brezil	100	150	96	4.24	4.57	4.61	0.90	1.00	0.38	0.63	0.02	0.05
brezil	200	150	99	7.92	8.26	8.27	0.75	1.00	0.41	0.56	0.06	2.39
brezil	300	150	98	11.51	11.92	11.95	0.80	2.00	0.42	0.61	3.09	200.08
brezil	400	150	99	15.10	15.45	15.46	0.75	1.00	0.42	0.58	1.21	100.22
brezil	500	150	95	18.76	19.10	19.17	0.75	3.00	0.42	0.53	7.83	300.48
brezil	600	150	82	22.32	22.61	22.79	0.75	1.00	0.42	0.56	21.69	170.00
eon	100	150	100	6.36	6.65	6.65	0.75	0.00	0.14	0.19	0.02	0.03
eon	200	150	100	11.54	11.77	11.77	0.75	0.00	0.15	0.17	0.06	0.11
eon	300	150	100	16.62	16.89	16.89	0.75	0.00	0.15	0.17	0.19	0.39
eon	400	150	100	21.47	21.85	21.85	0.75	0.00	0.15	0.17	0.57	1.58
eon	500	150	87	26.43	26.62	26.75	0.75	1.00	0.15	0.22	15.32	102.84
eon	600	150	42	31.36	31.63	32.24	0.75	2.00	0.15	0.19	66.10	202.98
mci	100	150	100	7.67	7.81	7.81	0.84	0.00	0.10	0.20	0.02	0.03
mci	200	150	100	13.42	13.58	13.58	0.80	0.00	0.10	0.13	0.08	0.14
mci	300	150	100	19.24	19.37	19.37	0.80	0.00	0.10	0.13	0.27	0.64
mci	400	150	97	25.00	25.14	25.17	0.80	1.00	0.10	0.13	4.15	100.77
mci	500	150	78	30.45	30.58	30.80	0.80	1.00	0.10	0.13	24.33	103.81
mci	600	150	33	36.00	36.11	36.87	0.80	2.00	0.11	0.20	76.84	204.42
nsf	100	150	99	7.97	8.38	8.39	0.90	1.00	0.05	0.08	0.09	6.55
nsf	200	150	100	15.06	15.45	15.45	0.75	0.00	0.05	0.06	0.10	0.22
nsf	300	150	100	21.96	22.29	22.29	0.75	0.00	0.06	0.08	0.48	1.70
nsf	400	150	90	28.81	29.18	29.28	0.75	1.00	0.06	0.06	11.46	110.38
nsf	500	150	41	35.79	36.13	36.81	0.79	2.00	0.06	0.16	70.70	218.00
nsf	600	150	23	42.52	42.94	43.93	0.75	3.00	0.06	0.09	104.04	301.59

resource model, the MIP model itself and the total number of frequencies required by the finite domain solver. The LP relaxation already is a very good approximation of the total cost, the MIP-LP gap never exceeds 0.90. The next column, *Max FD Gap*, shows the largest gap between MIP and FD solution, i.e. the number of frequencies added due to infeasibility or time out of the graph coloring model. This value never exceeds 3 in the examples shown. We then show average and maximal run times for the first and second phases of the decomposition on a Windows XP laptop with a 2.4GHz processor and 2GB of memory. Results were obtained using ECLiPSe 6.0 [16] with the eplex library [11] for the Coin-OR [9] CLP/CBC MIP solver.

Table 7 show corresponding results for the SAT model using minisat 1.14 [3], with a timeout of 100 seconds for each instance and each tested upper bound of the domain. If a timeout occurs, the problem is re-run adding frequencies until a solution is found within the timeout period. For increasing problem sizes the number of optimal solutions decreases sharply, in contrast to the finite domain model, while execution times are increasing significantly.

The hybrid model using the finite domain model is able to deal with much larger number of demands, as Table 8 shows. We consider the brezil network, and increase the number of demands up to 2000. The solving time for the first phase is not affected, as the model is not dependent on the number of demands, they only affect the upper bound of the domains T_s and the size of the coefficients P_{sd}. In the second phase the number of variables increases with the number of demands, and the Alldifferent constraints operate on larger number of variables, but the number of constraints is given by the topology and does not change.

Table 8. Increasing Demand Number (Extended Problem, 100 Runs Each)

Network	Dem.	λ	Opt.	Avg LP	Avg MIP	Avg FD	Max LP Gap	Max FD Gap	Avg MIP Time	Max MIP Time	Avg FD Time	Max FD Time
brezil	700	150	97	25.69	26.06	26.13	0.75	3.00	0.51	0.64	1.83	60.59
brezil	800	150	96	29.34	29.66	29.72	0.75	3.00	0.50	0.59	1.42	60.95
brezil	900	150	98	32.81	33.14	33.17	0.75	2.00	0.50	0.61	1.30	31.36
brezil	1000	150	99	36.34	36.68	36.69	0.75	1.00	0.50	0.63	1.24	2.13
brezil	1100	150	99	39.80	40.16	40.17	0.75	1.00	0.50	0.63	1.49	2.20
brezil	1200	150	99	43.28	43.61	43.62	0.75	1.00	0.50	0.63	2.24	46.16
brezil	1300	150	98	46.54	46.89	46.94	0.75	3.00	0.50	0.61	3.03	64.45
brezil	1400	150	99	49.85	50.21	50.23	0.75	2.00	0.50	0.63	2.79	33.95
brezil	1500	150	99	53.46	53.87	53.89	0.75	2.00	0.50	0.61	3.18	34.47
brezil	1600	150	98	56.95	57.28	57.30	0.75	1.00	0.50	0.59	4.49	72.05
brezil	1700	150	99	60.33	60.65	60.66	0.75	1.00	0.51	0.64	3.61	8.92
brezil	1800	150	99	63.93	64.25	64.26	0.75	1.00	0.51	0.61	4.08	9.49
brezil	1900	150	100	67.41	67.77	67.77	0.75	0.00	0.50	0.61	4.73	10.48
brezil	2000	150	99	70.83	71.09	71.10	0.75	1.00	0.51	0.66	6.05	94.73

Table 9. Increasing Network Size (Extended Problem, 100 Runs Each)

Network	Dem.	λ	Opt.	Avg LP	Avg MIP	Avg FD	Max LP Gap	Max FD Gap	Avg MIP Time	Max MIP Time	Avg FD Time	Max FD Time
r30	500	30	100	7.81	8.12	8.12	0.97	0.00	1.73	5.92	0.16	0.27
r40	500	30	100	4.14	4.52	4.52	0.92	0.00	12.42	177.45	0.13	0.19
r50	500	30	97	2.39	2.88	2.91	0.95	1.00	77.35	696.73	0.11	0.14
r60	500	30	100	1.57	2.05	2.05	0.86	0.00	127.75	245.25	0.10	0.13

The model is much more dependent on the size of the network. We consider in Table 9 random networks with 30-60 nodes, with a 0.25 probability for a link between two nodes. The times for the MIP (first) part of the model increase quickly with network size, and soon dominate the total execution times, while the second phase is barely affected. It is interesting that the execution times increase much more rapidly for the static design problem than for the demand acceptance problem discussed in [13], where network sizes up to 100 nodes can be solved within 30 seconds using the same environment.

5 Summary

In this paper we have considered some variants of the routing and wavelength assignment problem for optical networks. For the static design problem two possible objective functions have been proposed in the literature: The basic problem minimizing the maximal number of frequencies required on any link, and the extended problem minimizing the total number of frequencies used in the network. A decomposition into a MIP based routing part with a graph coloring second phase works very well in producing high quality solutions. Both the LP relaxation and the MIP solution of the first phase produce very accurate lower bounds on the total cost. The graph coloring problem in the basic model can be solved successfully by either MIP, SAT or finite domain constraint programming, constraint programming is slightly faster than SAT on the problem instances considered, and significantly faster than MIP. For the extended problem, constraint programming is much more stable and significantly faster than any of the competing methods.

Together with the results in [13] this shows that a decomposition of the RWA problem into MIP and FD phases can be highly successful, producing proven optimal or near-optimal results for a large set of problem instances.

References

1. Banerjee, D., Mukherjee, B.: A practical approach for routing and wavelength assignment in large wavelength-routed optical networks. IEEE Journal on Selected Areas in Communications 14(5), 903–908 (1996)
2. Bessiere, C., Hebrard, E., Hnich, B., Kiziltan, Z., Walsh, T.: Filtering algorithms for the nvalue constraint. Constraints 11(4), 271–293 (2006)
3. Eén, N., Sörensson, N.: An extensible SAT-solver. In: Giunchiglia, E., Tacchella, A. (eds.) SAT 2003. LNCS, vol. 2919, pp. 502–518. Springer, Heidelberg (2004)
4. Jaumard, B., Meyer, C., Thiongane, B.: ILP formulations for the routing and wavelength assignment problem: Symmetric systems. In: Resende, M., Pardalos, P. (eds.) Handbook of Optimization in Telecommunications, pp. 637–677. Springer, Heidelberg (2006)
5. Jaumard, B., Meyer, C., Thiongane, B.: Comparison of ILP formulations for the RWA problem. Optical Switching and Networking 4(3-4), 157–172 (2007)
6. Jaumard, B., Meyer, C., Thiongane, B.: On column generation formulations for the RWA problem. Discrete Applied Mathematics 157, 1291–1308 (2009)
7. Lever, J.: A local search/constraint propagation hybrid for a network routing problem. International Journal on Artificial Intelligence Tools 14(1-2), 43–60 (2005)
8. Liatsos, V., Novello, S., El Sakkout, H.: A probe backtrack search algorithm for network routing. In: Proceedings of the Third International Workshop on Cooperative Solvers in Constraint Programming, CoSolv 2003, Kinsale, Ireland (September 2003)
9. Lougee-Heimer, R.: The common optimization interface for operations research. IBM Journal of Research and Development 47, 57–66 (2003)
10. Ramaswami, R., Sivarajan, K.N.: Routing and wavelength assignment in all-optical networks. IEEE/ACM Trans. Netw. 3(5), 489–500 (1995)
11. Shen, K., Schimpf, J.: Eplex: Harnessing mathematical programming solvers for constraint logic programming. In: van Beek, P. (ed.) CP 2005. LNCS, vol. 3709, pp. 622–636. Springer, Heidelberg (2005)
12. Simonis, H.: Constraint applications in networks. In: Rossi, F., van Beek, P., Walsh, T. (eds.) Handbook of Constraint Programming. Elsevier, Amsterdam (2006)
13. Simonis, H.: A hybrid constraint model for the routing and wavelength assignment problem. In: Gent, I.P. (ed.) CP 2009. LNCS, vol. 5732, pp. 104–118. Springer, Heidelberg (2009)
14. Smith, B.M.: Symmetry and search in a network design problem. In: Barták, R., Milano, M. (eds.) CPAIOR 2005. LNCS, vol. 3524, pp. 336–350. Springer, Heidelberg (2005)
15. van Hoeve, W.J.: The alldifferent constraint: A survey. CoRR, cs.PL/0105015 (2001)
16. Wallace, M., Novello, S., Schimpf, J.: ECLiPSe: A platform for constraint logic programming. ICL Systems Journal 12(1) (May 1997)
17. Zang, H., Jue, J.P., Mukherjee, B.: A review of routing and wavelength assignment approaches for wavelength-routed optical WDM networks. Optical Networks Magazine, 47–60 (January 2000)

A Resource Cost Aware Cumulative

Helmut Simonis and Tarik Hadzic*

Cork Constraint Computation Centre
Department of Computer Science, University College Cork, Ireland
{h.simonis,t.hadzic}@4c.ucc.ie

Abstract. We motivate and introduce an extension of the well-known cumulative constraint which deals with time and volume dependent cost of resources. Our research is primarily interested in scheduling problems under time and volume variable electricity costs, but the constraint equally applies to manpower scheduling when hourly rates differ over time and/or extra personnel incur higher hourly rates. We present a number of possible lower bounds on the cost, including a min-cost flow, different LP and MIP models, as well as greedy algorithms, and provide a theoretical and experimental comparison of the different methods.

1 Introduction

The cumulative constraint [1,2] has long been a key global constraint allowing the effective modeling and resolution of complex scheduling problems with constraint programming. However, it is not adequate to handle problems where resource costs change with time and use, and thus must be considered as part of the scheduling. This problem increasingly arises with electricity cost, where time variable tariffs become more and more widespread. With the prices for electricity rising globally, the contribution of energy costs to total manufacturing cost is steadily increasing, thus making an effective solution of this problem more and more important. Figure 1 gives an example of the whole-sale price in the Irish electricity market for a sample day, in half hour intervals. Note that the range of prices varies by more than a factor of three, and the largest difference between two consecutive half-hour time periods is more than 50 units. Hence, large cost savings might be possible if the schedule of electricity-consuming activities takes that cost into account.

The problem of time and volume dependent resource cost equally arises in manpower scheduling, where hourly rates can depend on time, and extra personnel typically incurs higher costs. We suggest extending the standard CP scheduling framework to variable resource cost scenarios by developing a cost-aware extension of the cumulative constraint, CumulativeCost. The new constraint should support both the standard feasibility reasoning as well as reasoning about the cost of the schedule. As a first step in this direction we formally describe the semantics of CumulativeCost and discuss a number of algorithms for producing *high quality lower-bounds* (used for bounding or pruning during optimization) for this constraint. The proposed algorithms

---

* This work was supported by Science Foundation Ireland (Grant Numbers 05/IN/I886 and 05/IN.1/I886 TIDA 09).

J. Larrosa and B. O'Sullivan (Eds.): CSCLP 2009, LNAI 6384, pp. 76–89, 2011.

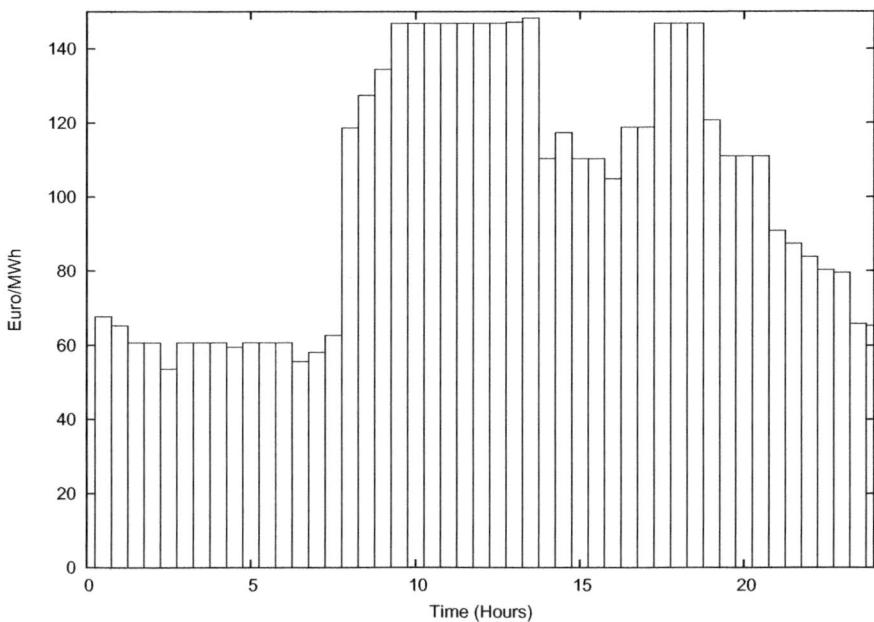

Fig. 1. Irish Electricity Price (Source: http://allislandmarket.com/)

are compared both theoretically and experimentally. Previous research on the cumulative constraint has focused along three lines: 1) improvement of the reasoning methods used inside the constraint propagation, see [10,7,8] for some recent results. 2) applying the constraint to new, sometimes surprising problem types, for example expressing producer/consumer constraints [9] or placement problems [4], and 3) extending the basic model to cover new features, like negative resource height, or non-rectangular task shapes [3,5]. The focus of this paper belongs to the last category.

2 The CumulativeCost Constraint

We start by formally defining our new constraint. It is an extension of the classical cumulative constraint [1]

$$\text{Cumulative}([s_1, s_2, ...s_n], [d_1, d_2, ...d_n], [r_1, r_2, ...r_n], l, p),$$

describing n tasks with start s_i, fixed duration d_i and resource use r_i, with an overall resource limit l and a scheduling period end p. Our new constraint is denoted as

$$\text{CumulativeCost}(\text{Areas}, \text{Tasks}, l, p, \text{cost}).$$

The *Areas* argument is a collection of m areas $\{A_1, \ldots, A_m\}$, which do not overlap and partition the entire available resource area $[0, p] \times [0, l]$. Each area A_j has a fixed position x_j, y_j, fixed width w_j and height h_j, and fixed per-unit cost c_j. Consider the running example in Fig. 2 (left). It has 5 areas, each of width 1 and height 3. Our definition allows that an area A_j could start above the bottom level ($y_j > 0$). This reflects the fact that the unit-cost does not only depend on the time of resource consumption but also on its volume. In our electricity example, in some environments, a limited amount of renewable energy may be available at low marginal cost, generated by wind-power or reclaimed process heat. Depending on the tariff, the electricity price may also be linked to the current consumption, enforcing higher values if an agreed limit is exceeded. We choose the numbering of the areas so that they are ordered by non-decreasing cost ($i \leq j \Rightarrow c_i \leq c_j$); in our example costs are $0, 1, 2, 3, 4$. There could be more than one area defined over the same time slot t (possibly spanning over other time-slots as well). If that is the case, we require that the area "above" has a higher cost. The electricity consumed over a certain volume threshold might cost more. The *Tasks* argument is a collection of n tasks. Each task T_i is described by its start s_i (between its earliest start $\underline{s_i}$ and latest start $\overline{s_i}$), and fixed duration d_i and resource use r_i. In our example, we have three tasks with durations $1, 2, 1$ and resource use $2, 2, 3$. The initial start times are $s_1 \in [2, 5], s_2 \in [1, 5], s_3 \in [0, 5]$. For a given task allocation, variable a_j states how many resource units of area A_j are used. For the optimal solution in our example we have $a_1 = 2, a_2 = 3, a_3 = 2, a_4 = 0, a_5 = 2$. Finally, we can define our constraint:

Definition 1. *Constraint* CumulativeCost *expresses the following relationships:*

$$\forall\, 0 \leq t < p : \quad pr_t := \sum_{\{i | s_i \leq t < s_i + d_i\}} r_i \leq l \tag{1}$$

$$\forall\, 1 \leq i \leq n : \quad 0 \leq \underline{s_i} \leq s_i < s_i + d_i \leq \overline{s_i} + d_i \leq p \tag{2}$$

$$ov(t, pr_t, A_j) := \begin{cases} \max(0, \min(y_j + h_j, pr_t) - y_j) & x_j \leq t < x_j + w_j \\ 0 & otherwise \end{cases} \tag{3}$$

$$\forall\, 1 \leq j \leq m : \quad a_j = \sum_{0 \leq t < p} ov(t, pr_t, A_j) \tag{4}$$

$$cost = \sum_{j=1}^{m} a_j c_j \tag{5}$$

For each time point t we first define the resource profile pr_t (the amount of resource consumed at time t). That profile must be below the overall resource limit l, as in the standard cumulative. The term $ov(t, pr_t, A_j)$ denotes the intersection of the profile at time t with area A_j, and the sum of all such intersections is the total resource usage a_j. The cost is computed by weighting each intersection a_j with the per-unit cost c_j of the area.

Note that our constraint is a strict generalization of the standard cumulative. Since enforcing generalized arc consistency (GAC) for Cumulative is NP-hard [6], the complexity of enforcing GAC over CumulativeCost is NP-hard as well. In the remainder of the paper we study the ability of CumulativeCost to reason about the cost through computation of lower bounds.

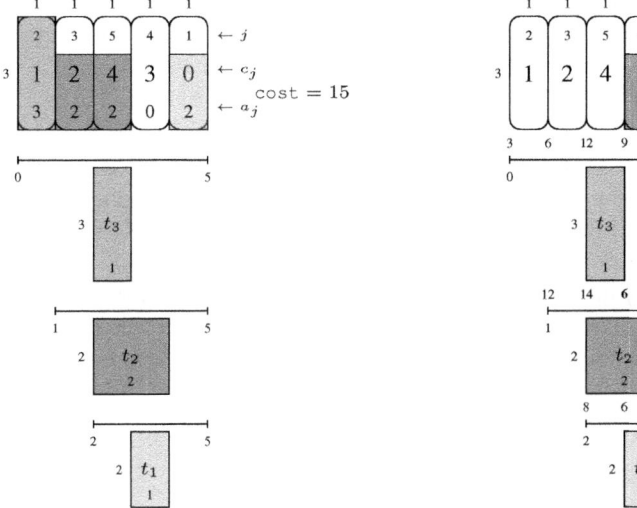

Problem and Optimal Solution   Element Model

Fig. 2. An example with 3 tasks and 5 areas. Areas are drawn as rounded rectangles at the top, the tasks to be placed below, each with a line indicating its earliest start and latest end. The optimal placement of tasks has cost 15 (left). The Element model produces a lower bound 6 (right).

3   Decomposition with Cumulative

We first present a number of models where we decompose the CumulativeCost into a standard Cumulative and some other constraint(s) which allow the computation of a lower bound for the cost variable. This cost component exchanges information with the Cumulative constraint only through domain restrictions on the s_i variables.

3.1   Element Model

The first approach is to decompose the constraint into a Cumulative constraint and a set of Element constraints. For a variable $x \in \{1, \ldots, n\}$ and a variable $y \in \{v_1, \ldots, v_n\}$, the element constraint $\text{element}(x, [v_1, v_2, ..., v_n], y)$ denotes that $y = v_x$. For each task i and start time t we precompute a cost of having a task T_i starting at time t in the cheapest area overlapping the time slot t (the bottom area). This value, denoted as v_{it}, is only a lower bound, and is incapable of expressing the cost if the task starts at t but in a higher (more expensive) area. A lower bound can then be expressed as

$$\text{lb} = \min \sum_{i=1}^{n} u_i$$

$$\forall\, 1 \leq i \leq n: \quad \text{element}(s_i, [v_{i1}, v_{i2}, ..., v_{ip}], u_i)$$

This lower bound can significantly underestimate the optimal cost since each task is assumed to be placed at its "cheapest" start time, even if the overall available volume would be exceeded. In our example, the v_{it} values for each start time are displayed above the time-lines for each task in Figure 2 (right). The lowest cost values for the tasks are 0, 6 and 0, resp., leading to a total lower bound estimate of 6.

3.2 Flow Model

We can also describe our problem as a min-cost flow model, where the flow cost provides a lower bound to the cost variable of our constraint. We need to move the flow $\sum_i d_i r_i$ from tasks to areas. Figure 3 shows the flow graph used, where the tasks T_i are linked to the areas A_j through flow variables f_{ij} which indicate how much volume of task i is contained in area j. The sum of all flows a_j through all areas must be equal to the total task volume. Only the links from A_j to the terminal T have non-zero cost c_j. The lower bound estimate is a min-cost flow.

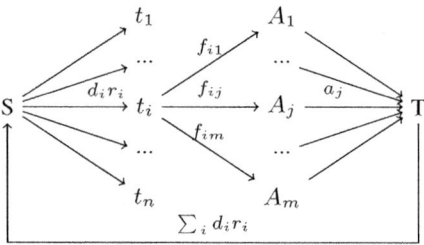

Fig. 3. Flow Graph

The model can also be expressed by a set of equations and inequalities.

$$\text{lb} = \min \sum_{j=1}^{m} a_j c_j \tag{6}$$

$$\forall\, 1 \le j \le m : \quad a_j = \sum_{i=1}^{n} f_{ij} \tag{7}$$

$$\forall\, 1 \le i \le n, \forall\, 1 \le j \le m : \quad \underline{f_{ij}} \le f_{ij} \le \overline{f_{ij}} \tag{8}$$

$$\forall\, 1 \le j \le m : \quad 0 \le \underline{a_j} \le a_j \le \overline{a_j} \le w_j h_j \tag{9}$$

$$\forall\, 1 \le i \le n : \quad \sum_{j=1}^{m} f_{ij} = d_i r_i \tag{10}$$

$$\forall\, 1 \le i \le n : \quad \sum_{i=1}^{n} d_i r_i = \sum_{j=1}^{m} a_j \tag{11}$$

While the flow model avoids placing too much work in the cheapest areas, it does allow splitting tasks over several, even non-contiguous areas to reduce overall cost, i.e. it ignores resource use and non-preemptiveness of tasks (Fig. 4, right).

The quality of the bound lb can be improved by computing tight upper bounds $\overline{f_{ij}}$. Given a task i with domain $d(s_i)$, we can compute the maximal overlap between the task and the area j as:

$$\overline{f_{ij}} = \max_{t \in d(s_i)} \max(0, (\min(x_j + w_j, t + d_i) - \max(x_j, t))) * \min(h_j, r_i) \qquad (12)$$

For the running example, the computed $\overline{f_{ij}}$ values are shown in Table 1 and the minimal cost flow is shown in Fig. 4. Its cost is $3*0 + 3*1 + 2*2 + 1*3 = 10$. Note how both tasks 1 and 2 are split into multiple pieces.

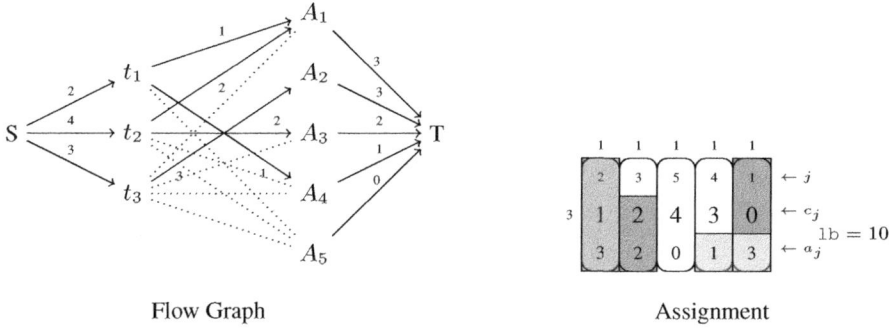

Flow Graph   Assignment

Fig. 4. Example Flow Bound

3.3 LP Models

The quality of the lower bound can be further improved by adding to the flow model some linear constraints, which limit how much of some task can be placed into low-cost areas. These models are no longer flows, but general LP models. We consider two variations denoted as *LP1* and *LP2*. We obtain *LP1* by adding equations (13) to the LP formulation of the flow model (Constraints 6-11).

$$\forall 1 \leq j \leq m: \quad \sum_{i=1}^{n} \sum_{k=1}^{j} f_{ik} = \sum_{k=1}^{j} a_k \leq \overline{B_j} = \sum_{i=1}^{n} \overline{b_{ij}} \qquad (13)$$

The $\overline{b_{ij}}$ values tell us how much of task i can be placed into the combination of cheaper areas $\{A_1, A_2, ...A_j\}$. Note that this is a stronger estimate, since $\overline{b_{ij}} \leq \sum_{k=1}^{j} \overline{f_{ij}}$. Therefore, the LP models dominate the flow model, but require a more expensive LP optimization at each step. $\overline{B_j}$ denotes the total amount of resources (from all tasks) that is possible to place into the first j areas. We can compute the $\overline{b_{ij}}$ values with a sweep along the time axis.

Table 1 shows the $\overline{f_{ij}}$, $\overline{b_{ij}}$ and $\overline{B_j}$ values for the running example. The flow solution in Fig. 4 does not satisfy equation (13) for $j = 3$, as $a_1 + a_2 + a_3 = 3 + 3 + 2 = 8 > b_{13} + b_{23} + b_{33} = 2 + 2 + 3 = 7$. Solving the LP model finds an improved bound of 11.

Table 1. $\overline{f_{ij}}$, $\overline{b_{ij}}$ and $\overline{B_j}$ Values for Example

$\overline{f_{ij}}$	1 2 3 4 5
1	2 0 0 2 2
2	2 0 2 2 2
3	3 3 3 3 3

$\overline{b_{ij}}$	1 2 3 4 5
1	2 2 2 2 2
2	2 2 2 4 4
3	3 3 3 3 3
$\overline{B_j}$	7 7 7 9 9

The *LP2 model* in principle may produce potentially even stronger bounds by replacing constraints (13) with more detailed constraints on individual $\overline{b_{ij}}$:

$$\forall\, 1 \leq i \leq n, \forall\, 1 \leq j \leq m : \quad \sum_{k=1}^{j} f_{ik} \leq \overline{b_{ij}} \tag{14}$$

In our example though, this model produces the same bound as model *LP1*.

4   Coarse Models and Greedy Algorithms

If we want to avoid running an LP solver inside our constraint propagator, we can derive other lower bounds based on greedy algorithms.

4.1   Model A

We can compute a lower bound through a coarser model (denoted as *Model A*) which has m variables u_j, denoting the total amount of work going into A_j.

$$\mathtt{lb} = \min \sum_{j=1}^{m} u_j c_j \tag{15}$$

$$\forall\, 1 \leq j \leq m : \quad u_j \leq w_j h_j \tag{16}$$

$$\forall\, 1 \leq j \leq m : \quad u_j \leq \sum_{i=1}^{n} \overline{f_{ij}} \tag{17}$$

$$\sum_{j=1}^{m} u_j = \sum_{i=1}^{n} d_i r_i \tag{18}$$

It can be shown that this bound can be achieved by Algorithm A, which can be described by the recursive equations:

$$\mathtt{lb} = \sum_{j=1}^{m} u_j c_j \tag{19}$$

$$\forall\, 1 \leq j \leq m : \quad u_j = \min(\sum_{i=1}^{n} d_i r_i - \sum_{k=1}^{j-1} u_k, \sum_{i=1}^{n} \overline{f_{ij}}, w_j h_j) \tag{20}$$

The algorithm tries to fill the cheapest areas first as far as possible. For each area we compute the minimum of the remaining unallocated task area, and the maximum amount that can be placed into the area based on the $\overline{f_{ij}}$ bounds and the overall area size.

4.2 Model B

We can form a stronger model (*Model B*) by extending Model A with the constraints:

$$\forall\, 1 \leq j \leq m: \quad \sum_{k=1}^{j} u_k \leq \overline{B_j} = \sum_{i=1}^{n} \overline{b_{ij}} \tag{21}$$

It can be shown that the lower-bound can be computed through Algorithm B, which extends Algorithm A by also considering the $\overline{B_j}$ bounds (constraints (13)) and is recursively defined as:

$$\mathtt{lb} = \sum_{j=1}^{m} u_j c_j \tag{22}$$

$$\forall\, 1 \leq j \leq m: \quad u_j = \min\Big(\sum_{i=1}^{n} d_i r_i - \sum_{k=1}^{j-1} u_k,\ \sum_{i=1}^{n} \overline{f_{ij}},\ w_j h_j,\ \sum_{i=1}^{n} \overline{b_{ij}} - \sum_{k=1}^{j-1} u_k\Big) \tag{23}$$

Both algorithms A and B only compute a bound on the cost, and do not produce a complete assignment of tasks to areas. On the other hand, they only require a single iteration over the areas, provided the $\overline{f_{ij}}$ and $\overline{b_{ij}}$ bounds are known. Figure 5 compares the execution of algorithms A and B on the running example (top) and shows the resulting area utilization (bottom). Algorithm A computes a bound of 9, which is weaker than the Flow Model ($\mathtt{lb} = 10$). Algorithm B produces 11, the same as models LP1 and LP2. Note that we can not easily determine which tasks are used to fill which area.

5   Direct Model

All previous models relied on a decomposition using a standard cumulative constraint to enforce overall resource limits. We now extend the model of the cumulative constraint given in [6] to handle the cost component directly (equations (24)-(33)).

We introduce binary variables y_{it} which state whether task i starts at time t. For each task, exactly one of these variables will be one (constraint (30)). Equations (29) connect the s_i and y_{it} variables. Continuous variables pr_t describe the resource profile at each time point t, all values must be below the resource limit l. The profile is used in two ways: In (31), the profile is built by cumulating all active tasks at each time-point. In (32), the profile overlaps all areas active at a time-point, where the contribution of area j at time-point t is called z_{jt} (a continuous variable ranging between zero and h_j). Adding all contributions of an area leads to the resource use a_j for area j. This model combines the start-time based model of the cumulative with a standard LP formulation of the convex, piece-wise linear cost of the resource profile at each time point. Note that this model relies on the objective function to fill up cheaper areas to capacity before using more expensive ones. Enforcing the integrality in (26) leads to a mixed integer programming model *DMIP*, relaxing the integrality constraint leads to the LP model *DLP*. The MIP model solves the cumulative-cost constraint to optimality, thus providing

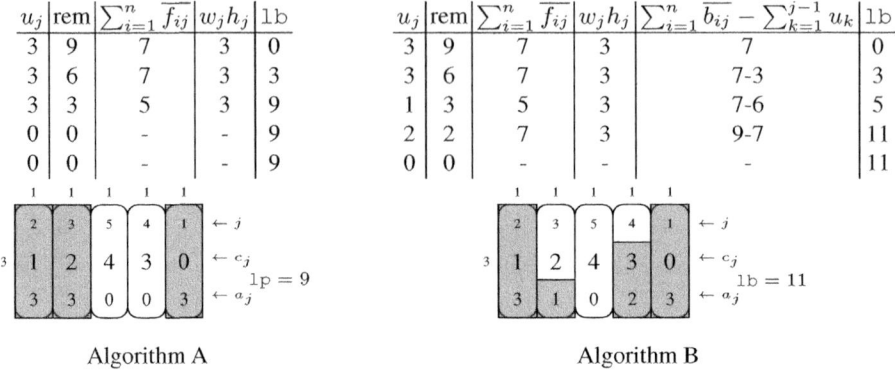

Algorithm A   Algorithm B

Fig. 5. Execution of algorithms A and B on the running example

an exact bound for the constraint. We can ignore the actual solution if we want to use the constraint in a larger constraint problem.

$$\mathtt{lb} = \min \sum_{j=1}^{m} a_j c_j \tag{24}$$

$$\forall\, 0 \le t < p: \quad pr_t \in [0, l] \tag{25}$$

$$\forall\, 1 \le i \le n, 0 \le t < p: \quad y_{it} \in \{0, 1\} \tag{26}$$

$$\forall\, 1 \le j \le m, \forall\, x_j \le t < x_j + w_j: \quad z_{jt} \in [0, h_j] \tag{27}$$

$$\forall\, 1 \le j \le m: \quad 0 \le \underline{a_j} \le a_j \le \overline{a_j} \le w_j h_j \tag{28}$$

$$\forall\, 1 \le i \le n: \quad s_i = \sum_{t=0}^{p-1} t y_{it} \tag{29}$$

$$\forall\, 1 \le i \le n: \quad \sum_{t=0}^{p-1} y_{it} = 1 \tag{30}$$

$$\forall\, 0 \le t < p: \quad pr_t = \sum_{t' \le t < t' + d_i} y_{it'} r_i \tag{31}$$

$$\forall\, 0 \le t < p: \quad pr_t = \sum_{j=1}^{m} z_{jt} \tag{32}$$

$$\forall\, 1 \le j \le m: \quad a_j = \sum_{t=x_j}^{x_j + w_j - 1} z_{jt} \tag{33}$$

Example. Figure 6 shows the solutions of *DLP* and *DMIP* on the running example. The linear model on the left uses fractional assignments for task 2, splitting it into two segments. This leads to a bound of 12, while the *DMIP* model computes a bound of 15.

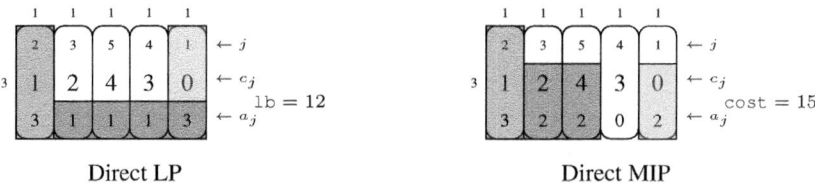

Direct LP	Direct MIP

Fig. 6. Direct Model Example

6 Comparison

We now want to compare the relative strength of the different algorithms. We first define the concept of strength in our context.

Definition 2. *We call algorithm q* `stronger` *than algorithm p if its computed lower bound* lb_q *is always greater than or equal to* lb_p, *and for some instances is strictly greater than.*

Theorem 1. *The relative strength of the algorithms can be shown as a diagram, where a line p → q indicates that algorithm q is stronger than algorithm p, and a dotted line between two algorithms states that they are incomparable.*

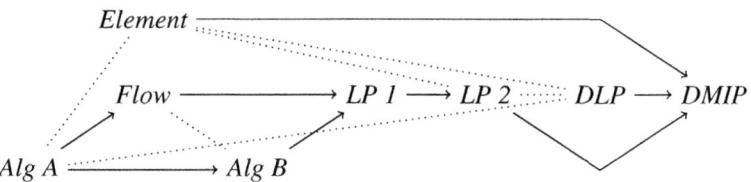

Proof. We need two lemmas, which are presented without proof:

Lemma 1. *Every solution of the Flow Model is a solution of Model A.*

Lemma 2. *Every solution of the model LP1 is a solution of Model B.*

This suffices to show that the Flow Model is stronger than Model A, and Model LP1 is stronger than Model B. The other "stronger" relations are immediate. All "incomparable" results follow from experimental outcomes. Figure 7 shows as example where the Element Model and Algorithm A outperform the model DLP. Table 4 below shows cases where the Element Model is stronger than Model A and Model LP2, and where the Flow Model outperforms Algorithm B.

Why are so many of the models presented incomparable? They all provide different relaxations of the basic constraints which make up the `CumulativeCost` constraint. In Table 2 we compare our methods based on four main concepts, whether the algorithms respect the capacity limits for each single area, or for (some of the) multiple areas, whether the earliest start/latest end limits for the start times are respected, and whether the tasks are placed as rectangles with the given width and height. Each method respects a different subset of the constraints. Methods are comparable if for all concepts one model is stronger than the other, but incomparable if they relax different concepts.

lb(Element)=2 > lb(DLP)=0
lb(Alg A)=2 > lb(DLP)=0

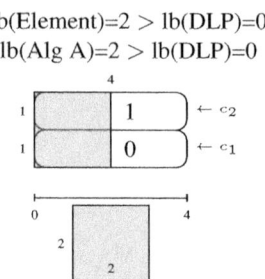

Fig. 7. Example: Element and Flow Models stronger than DLP

Table 2. Model Satisfies Basic Constraints/Complexity

Method	Single Area Capacity	Multi Area Capacity	Earliest Start Latest End	Task Profile	Model Setup	Variables	Constraints
Element	no	no	yes	yes	$\mathcal{O}(npm)$	-	-
Flow	yes	no	yes	$\leq$ height	$\mathcal{O}(nm)$	$\mathcal{O}(nm)$	$\mathcal{O}(n+m)$
LP 1	yes	$\overline{B_j}$	yes	$\leq$ height	$\mathcal{O}(nm^2)$	$\mathcal{O}(nm)$	$\mathcal{O}(n+m)$
LP 2	yes	$\overline{b_{ij}}$	yes	$\leq$ height	$\mathcal{O}(nm^2)$	$\mathcal{O}(nm)$	$\mathcal{O}(nm)$
Alg A	yes	no	no	no	$\mathcal{O}(nm)$	-	-
Alg B	yes	$\overline{B_j}$	no	no	$\mathcal{O}(nm^2)$	-	-
DLP	yes	yes	yes	width		$\mathcal{O}(np+mp)$	$\mathcal{O}(n+m+p)$
DMIP	yes	yes	yes	yes		$\mathcal{O}(np+mp)$	$\mathcal{O}(n+m+p)$

Without space for a detailed complexity analysis, we just list on the right side of Table 2 the main worst-case results for each method. We differentiate the time required to preprocess and generate the constraints, and the number of variables and constraints for the LP and MIP models. Recall that n is the number of tasks, m the number of areas and p the length of the scheduling period. Note that these numbers do not estimate the time for computing the lower bounds but for setting-up the models.

7   Experiments

We implemented all algorithms discussed in the paper, referring to them as {*DMIP, Element, Algo A, Algo B, Flow, LP1, LP2, DLP*}. For convenience we denote them also as $Algo_0, \ldots, Algo_7$. We evaluated their performance over a number of instances, denoted E. Each instance $e \in E$ is defined by the areas $A_1, \ldots, A_m$ and tasks $T_1, \ldots, T_n$. In all experiments we use a day-long time horizon divided into 48 half-hour areas and two cost settings. In the first setting for each half-hour period we fix a price to given values from a real-world price distribution (Fig. 1) over all the task specifications. We denote this as a *fixed* cost distribution ($cost = fixed$). Alternatively, for each area we choose a random cost from interval $[0, 100]$, for each task specification. We refer to this as a *random* cost distribution ($cost = random$). In both cases the costs are normalized by subtracting the minimal cost $c_{min} = min_{j=1}^m c_j$ from each area cost. For a given $n, d_{max}, r_{max}, \Delta$ we generate tasks $T_1, \ldots, T_n$ with randomly selected durations $d_i \in [1, d_{max}]$ and resource consumptions $r_i \in [1, r_{max}]$, $i = 1, \ldots, n$. Δ denotes the maximal distance between the earliest and latest start time. For a given Δ we randomly select the interval $s_i \in [\underline{s_i}, \overline{s_i}]$ so that $0 \leq \overline{s_i} - \underline{s_i} \leq \Delta$. One of the major parameters we consider is *utilization* defined as a portion of the total available area that must be covered by the tasks: $util = \sum_{i=1}^{n} d_i r_i / (l \cdot p)$. In the experiments we first decide the desired level of utilization, $util$, and then select a corresponding capacity limit $l = \lceil \frac{\sum_{i=1}^{n} d_i r_i}{util \cdot p} \rceil$. Hence, each instance scenario is obtained by specifying $(n, d_{max}, r_{max}, \Delta, util, cost)$.

We used *Choco v2.1.1*[1] as the underlying constraint solver, and *CPLEX v12.1*[2] as a MIP/LP solver. The experiments ran as a single thread on a Dual Quad Core Xeon CPU, 2.66GHz with 12MB of L2 cache per processor and 16GB of RAM overall, running Linux 2.6.25 x64. All the reported times are in milliseconds.

General Comparison. In the first set of experiments we compare the quality of the bounds for all algorithms presented. Let val_i denote a value (lower bound) returned by $Algo_i$. The *DMIP* algorithm, $Algo_0$ solves the problem exactly, yielding the value val_0 equal to the optimum. For other algorithms i we define its quality q_i as val_i / val_0.

For a given instance we say that $Algo_i$ is *optimal* if $q_i = 100\%$. For all algorithms except *DMIP* (i.e. $i > 0$), we say that they are the *best* if they have the maximal lower bound over the non-exact algorithms, i.e. $q_i = max_{j>0} q_j$. Note that the algorithm might be best but not optimal. For a given scenario $(n, d_{max}, r_{max}, \Delta, util, cost)$ we generate 100 instances, and compute the following values: 1) number of times that the algorithm was optimal or best 2) average and minimal quality q_i, 3) average and maximal execution time. We evaluate the algorithms over scenarios with $n = 100$ tasks, $d_{max} = 8$, $r_{max} = 8$, $\Delta = 10$, $util \in \{30\%, 50\%, 70\%, 80\%\}$ and $cost \in \{fixed, random\}$. This leads in total to eight instance scenarios presented in Table 3. The first column indicates the scenario (utilization and cost distribution). The last eight columns indicate the values described above for each algorithm. First note that the bound estimates for algorithms *B, LP1, LP2* are on average consistently higher than 95% over all scenarios, but that *DLP* provides exceptionally high bounds, on average over 99.6% for each scenario. While algorithms *A* and *Flow* can provide weaker bounds for lower utilization, the bounds improve for higher utilization. Algorithm *Element* on the other hand performs better for lower utilization (since its bound ignores capacity l) and deteriorates with higher utilization.

We also performed experiments over unrestricted initial domains, i.e. where $s_i \in [0, p - 1]$. While most of the algorithms improved their bound estimation, *Element* performed much worse for the fixed cost distribution, reaching quality as low as 12% for high utilization. On the other hand, reducing Δ to 5 significantly improved the quality of Element, while slightly weakening the quality of other algorithms. In all the cases, the worst-case quality of *DLP* was still higher than 98.6%.

Aggregate Pairwise Comparison. In Table 4 we compare all algorithms over the entire set of instances that were generated under varying scenarios involving 100 tasks. For

---

Table 3. Evaluation of algorithms for $d_{max} = r_{max} = 8$ and $\Delta = 10$ with varying *util* and *cost*

Scenario	Key	DMIP		Element		A		B		Flow		LP1		LP2		DLP	
util=30	Opt/Best	100	-	92	92	0	0	0	0	0	0	0	0	0	0	93	100
fixed	Avg/Min Quality	100.0	100.0	99.998	99.876	57.268	33.055	99.77	99.337	97.429	94.665	99.77	99.337	99.77	99.337	99.999	99.994
	Avg/Max Time	29	194	185	509	8	73	12	126	34	138	150	277	211	617	111	380
util=50	Opt/Best	100	-	2	2	0	0	0	0	0	0	0	0	0	0	7	100
fixed	Avg/Min Quality	100.0	100.0	99.038	94.641	68.789	54.22	99.131	95.963	97.816	95.89	99.407	97.773	99.435	97.913	99.948	99.358
	Avg/Max Time	89	2,108	176	243	6	12	7	94	33	130	139	274	194	275	96	181
util=70	Opt/Best	100	-	0	0	0	0	0	0	0	1	0	0	0	5	1	100
fixed	Avg/Min Quality	100.0	100.0	93.541	81.603	84.572	69.953	96.495	87.884	99.1	96.994	99.24	97.838	99.346	98.071	99.764	98.992
	Avg/Max Time	2,107	32,666	177	242	7	103	8	72	34	97	136	239	213	1,798	110	1,551
util=80	Opt/Best	100	-	0	0	0	0	0	1	0	4	0	20	0	21	0	100
fixed	Avg/Min Quality	100.0	100.0	88.561	70.901	92.633	81.302	96.163	89.437	99.34	96.728	99.354	96.737	99.392	96.737	99.649	98.528
	Avg/Max Time	13,017	246,762	206	450	10	67	15	96	38	124	156	235	220	426	125	299
util=30	Opt/Best	100	-	94	94	0	0	0	0	0	0	0	0	0	0	97	100
random	Avg/Min Quality	100.0	100.0	99.996	99.872	58.094	41.953	96.965	93.237	73.759	54.641	96.965	93.237	96.966	93.254	99.999	99.977
	Avg/Max Time	29	154	192	427	7	24	8	42	32	94	145	224	203	361	99	274
util=50	Opt/Best	100	-	0	0	0	0	2	8	2	8	2	8	2	8	5	100
random	Avg/Min Quality	100.0	100.0	88.277	30.379	76.457	57.049	96.585	92.563	83.314	69.178	96.619	92.604	96.861	93.242	99.93	99.724
	Avg/Max Time	2,452	62,168	202	814	10	99	13	131	43	177	165	380	238	903	108	327
util=70	Opt/Best	100	-	0	0	0	0	0	0	0	0	0	0	0	0	0	100
random	Avg/Min Quality	100.0	100.0	91.045	72.06	89.784	75.822	95.242	90.496	92.953	84.277	95.953	92.24	96.947	94.012	99.697	99.374
	Avg/Max Time	72,438	2,719,666	226	436	13	74	26	98	70	178	223	543	280	566	108	428
util=80	Opt/Best	100	-	0	0	0	0	0	0	0	0	0	0	0	0	0	100
random	Avg/Min Quality	100.0	100.0	86.377	72.566	94.813	88.039	96.092	89.233	97.231	93.919	97.658	94.917	98.426	96.342	99.626	99.161
	Avg/Max Time	684,660	8,121,775	320	2,148	16	100	31	180	63	370	223	1,586	363	23,91	286	7,242

any two algorithms $Algo_i$ and $Algo_j$ let $E_{ij} = \{e \in E \mid q_{ei} > q_{ej}\}$ denote the set of instances where $Algo_i$ outperforms $Algo_j$. Let $num_{ij} = |E_{ij}|$ denote the number of such instances, while avg_{ij} and max_{ij} denote the average and maximal difference in quality over E_{ij}. In the (i,j)-th cell of Table 4 we show $(num_{ij}, avg_{ij}, max_{ij})$. In approximately 2000 instances *DLP* strictly outperforms all other non-exact algorithms in more than 1700 cases and is never outperformed by another. Algorithm *B* outperforms *Flow* (1107 cases) more often than *Flow* outperforming *B* (333 cases). Furthermore, *LP2* outperforms *LP1* in about 700 cases. Interestingly, even though *Element* produces on average weaker bounds, it is able to outperform all non-exact algorithms except *DLP* on some instances.

Table 4. Comparison of algorithms over all instances generated for experiments with $n = 100$ tasks

	Element	A	B	Flow	LP1	LP2	DLP
DMIP	1802 26.39 88.62	1944 17.47 68.55	1944 3.99 30.71	1944 9.85 68.55	1944 3.49 30.71	1944 3.24 30.71	1387 0.18 1.47
EL	-	1034 25.62 66.95	656 3.0 22.07	856 15.65 65.82	650 3.01 22.07	621 3.04 22.07	-
A	1052 38.1 88.61	-	1439 18.22 66.65				
B	1429 29.22 88.61	1439 18.22 66.65	-	1107 11.02 51.86			
FLW	1230 33.97 88.61	1184 12.51 64.35	333 2.39 10.49	-			
LP1	1436 29.74 88.61	1441 18.86 66.65	726 1.33 10.51	1413 8.75 51.86	-		
LP2	1465 29.44 88.61	1441 19.19 66.65	846 1.71 10.64	1425 9.02 51.86	690 0.7 5.09	-	
DLP	1802 26.24 88.61	1752 19.24 68.55	1751 4.28 30.71	1747 10.82 68.55	1727 3.78 30.71	1725 3.51 30.71	-

Varying Number of Tasks. Finally, we evaluated the effect of increasing the number of tasks. We considered scenarios $n \in \{50, 100, 200, 400\}$ where $d_{max} = 8$, $r_{max} = 8$, $\Delta = 10$, $util = 70$ and $cost = random$. The results are reported in Table 5. We can notice that all algorithms improve the quality of their bounds with an increase in the number of tasks. While the time to compute the bounds grows for the non-exact algorithms, interestingly this is not always the case for the *DMIP* model. The average time to compute the optimal value peaks for $n = 100$ (72 seconds on average) and then reduces to 24 and 9 seconds for $n = 200$ and $n = 400$ respectively.

Table 5. Evaluation of algorithms for random cost distribution, $\Delta = 10$ and $util = 70$

Scenario	Key	DMIP		Element		A		B		Flow		LP1		LP2		DLP	
n=50	Opt/Best	100	0	0	0	0	0	0	0	0	0	0	0	0	0	0	100
	Avg/Min Quality	100.0	100.0	89.331	71.865	88.833	70.372	93.991	86.143	92.384	84.589	94.863	87.417	95.899	89.937	98.664	95.523
	Avg/Max Time	1,334	31,426	102	314	8	83	6	36	23	78	93	254	132	298	75	233
n=100	Opt/Best	100	0	0	0	0	0	0	0	0	0	0	0	0	0	0	100
	Avg/Min Quality	100.0	100.0	91.045	72.06	89.784	75.822	95.242	90.496	92.953	84.277	95.953	92.24	96.947	94.012	99.697	99.374
	Avg/Max Time	72,438	2,719,666	226	436	13	74	26	98	70	178	223	543	280	566	152	428
n=200	Opt/Best	100	0	0	0	0	0	0	0	0	0	0	0	0	0	0	100
	Avg/Min Quality	100.0	100.0	92.537	84.06	89.819	79.862	95.885	93.01	92.566	83.516	96.48	93.328	97.158	93.885	99.936	99.833
	Avg/Max Time	24,491	233,349	395	700	19	148	22	113	83	208	341	533	468	638	226	456
n=400	Opt/Best	100	0	0	0	0	0	0	0	0	0	0	0	0	0	0	100
	Avg/Min Quality	100.0	100.0	93.239	86.419	90.417	84.205	96.3	92.721	92.939	86.658	96.716	93.275	97.23	95.013	99.985	99.961
	Avg/Max Time	9,379	158,564	831	1,222	31	164	36	189	181	305	923	3,053	1,214	3,434	484	871

8   Conclusions and Future Work

We have introduced the `CumulativeCost` constraint, a resource cost-aware extension of the standard cumulative constraint, and suggested a number of algorithms to compute lower bounds on its cost. We compared the algorithms both theoretically and experimentally and discovered that while most of the approaches are mutually incomparable, the *DLP* model clearly dominates all other algorithms for the experiments considered. Deriving lower bounds is a necessary first step towards the development of the `CumulativeCost` constraint. We plan to evaluate the performance of the constraint in a number of real-life scheduling problems, which first requires development of domain filtering algorithms and specialized variable/value ordering heuristics.

References

1. Aggoun, A., Beldiceanu, N.: Extending CHIP in order to solve complex scheduling problems. Journal of Mathematical and Computer Modelling 17(7), 57–73 (1993)
2. Baptiste, P., Le Pape, C., Nuijten, W.: Constraint-Based Scheduling: Applying Constraint Programming to Scheduling Problems. Kluwer, Dordrecht (2001)
3. Beldiceanu, N., Carlsson, M.: A new multi-resource cumulatives constraint with negative heights. In: Van Hentenryck, P. (ed.) CP 2002. LNCS, vol. 2470, pp. 63–79. Springer, Heidelberg (2002)
4. Beldiceanu, N., Carlsson, M., Poder, E.: New filtering for the cumulative constraint in the context of non-overlapping rectangles. In: Perron, L., Trick, M.A. (eds.) CPAIOR 2008. LNCS, vol. 5015, pp. 21–35. Springer, Heidelberg (2008)
5. Beldiceanu, N., Poder, E.: A continuous multi-resources cumulative constraint with positive-negative resource consumption-production. In: Van Hentenryck, P., Wolsey, L.A. (eds.) CPAIOR 2007. LNCS, vol. 4510, pp. 214–228. Springer, Heidelberg (2007)
6. Hooker, J.: Integrated Methods for Optimization. Springer, New York (2007)
7. Mercier, L., Van Hentenryck, P.: Edge finding for cumulative scheduling. INFORMS Journal on Computing 20(1), 143–153 (2008)
8. Schutt, A., Feydy, T., Stuckey, P.J., Wallace, M.: Why cumulative decomposition is not as bad as it sounds. In: Gent, I.P. (ed.) CP 2009. LNCS, vol. 5732, pp. 746–761. Springer, Heidelberg (2009)
9. Simonis, H., Cornelissens, T.: Modelling producer/consumer constraints. In: Montanari, U., Rossi, F. (eds.) CP 1995. LNCS, vol. 976, pp. 449–462. Springer, Heidelberg (1995)
10. Vilím, P.: Max energy filtering algorithm for discrete cumulative resources. In: van Hoeve, W.-J., Hooker, J.N. (eds.) CPAIOR 2009. LNCS, vol. 5547, pp. 294–308. Springer, Heidelberg (2009)

Integrating Strong Local Consistencies into Constraint Solvers[*]

Julien Vion[1,2], Thierry Petit[3], and Narendra Jussien[3]

[1] Univ. Lille Nord de France, F-59500 Lille, France
[2] UVHC, LAMIH FRE CNRS 3304, F-59313 Valenciennes, France
julien.vion@univ-valenciennes.fr
[3] École des Mines de Nantes,
LINA UMR CNRS 6241,
4, rue Alfred Kastler, 44307 Nantes, France
thierry.petit@mines-nantes.fr, narendra.jussien@mines-nantes.fr

Abstract. This article presents a generic scheme for adding strong local consistencies to the set of features of constraint solvers, which is notably applicable to event-based constraint solvers. We encapsulate a subset of constraints into a global constraint. This approach allows a solver to use different levels of consistency for different subsets of constraints in the same model. Moreover, we show how strong consistencies can be applied with different kinds of constraints, including user-defined constraints. We experiment our technique with a coarse-grained algorithm for Max-RPC, called Max-RPC^{rm}, and a variant of it, L-Max-RPC^{rm}. Experiments confirm the interest of strong consistencies for Constraint Programming tools.

1 Introduction

This paper presents a generic framework for integrating strong local consistencies into Constraint Programming (CP) tools, especially event-based solvers. It is successfully experimented using Max-RPC^{rm} and L-Max-RPC^{rm}, recent coarse-grained algorithms for Max-RPC and a variant of this consistency [25].

The most successful techniques for solving problems with CP are based on local consistencies. Local consistencies remove values or assignments that cannot belong to a solution. To enforce a given level of local consistency, *propagators* are associated with constraints. A propagator is complete when it eliminates all the values that cannot satisfy the constraint. One of the reasons for which CP is currently applied with success to real-world problems is that some propagators are encoded through *filtering algorithms*, which exploit the semantics of the constraints. Filtering algorithms are often derived from well-known Operations Research techniques. This provides powerful implementations of propagators.

---

[*] This work was supported by the ANR French research funding agency, through the CANAR project (ANR-06-BLAN-0383-03).

J. Larrosa and B. O'Sullivan (Eds.): CSCLP 2009, LNAI 6384, pp. 90–104, 2011.

Many solvers use an AC-5 based propagation scheme [23]. We call them event-based solvers. Each propagator is called according to the events that occur in the domains of the variables involved in its constraint. For instance, an event may be a value deleted by another constraint. At each node of the search tree, the pruning is performed within the constraints. The fixed point is obtained by propagating events among all the constraints. In this context, generalized arc-consistency (GAC) is, *a priori*, the highest level of local consistency that can be enforced (all propagators are complete).

On the other hand, local consistencies that are stronger than GAC [9,6] require to take into account several constraints at a time in order to be enforced. Therefore, it is considered that such strong consistencies cannot easily be integrated into CP toolkits, especially event-based solvers. Toolkits do not feature those consistencies,[1] and they are not used for solving real-life problems.

This article demonstrates that strong local consistencies are wrongly excluded from CP tools. We present a new generic paradigm to add strong local consistencies to the set of features of constraint solvers. Our idea is to define a *global constraint* [7,1,19], which encapsulates a subset of constraints of the model. The strong consistency is enforced on this subset of constraints. Usually, a global constraint represents a sub-problem with fixed semantics. It is not the case for our global constraint: it is used to apply a propagation technique on a given subset of constraints, as it was done in [20] in the context of over-constrained problems. Our scheme may be connected to Bessière & Régin's "on the fly" sub-problem solving [5]. However, there is a fundamental divergence as our scheme is aimed at encoding strong consistencies. Thus, we keep a local evaluation of the supports for each constraint in the encapsulated model.

This approach provides some new possibilities compared with the state of the art. A first improvement is the ability to use different levels of consistency for different subsets of constraints in the same constraint model. This feature is an alternative to the heuristics for dynamically switching between different levels of consistency during search [21]. A second one is to apply strong consistencies to all kinds of constraints, including user-defined constraints or arithmetic expressions. Finally, within the global constraint, it is possible to define any strategy for handling events. One may order events variable per variable instead of considering successively each encapsulated constraint. Event-based solvers generally do not provide such a level of precision.

We experiment our framework with the Max-RPC strong consistency [8], using the Choco CP solver [15]. We use a coarse-grained algorithm for Max-RPC, called Max-RPC^{rm} [25]. This algorithm exploits backtrack-stable data structures in a similar way to AC-3^{rm} [17].

Section 2 presents the background about constraint networks and local consistencies useful to understand our contributions. Section 3 presents the generic integration scheme and it specialization to specific strong local consistencies.

---

[1] Some strong consistencies such as SAC [3] can be implemented using assignment and propagation methods, and some solvers may feature such ones.

Section 4 describes Max-RPC^{rm} and L-Max-RPC^{rm}. Section 5 details the experimental evaluation of our work. Finally, we discuss the perspectives and we conclude.

2   Background

A *constraint network* $\mathcal{N}$ is a triple $(\mathcal{X}, \mathcal{D}, \mathcal{C})$ which consists of :

- a set of n variables $\mathcal{X}$,
- a set of domains $\mathcal{D}$, where the domain $\mathrm{dom}(X) \in \mathcal{D}$ of the variable X is the finite set of at most d values that the variable X can take, and
- a set $\mathcal{C}$ of e constraints that specify the allowed combinations of values for given subsets of variables.

A variable/value couple (X, v) will be denoted X_v. An *instantiation* I is a set of variable/values couples. I is *valid* iff for any variable X involved in I, $v \in \mathrm{dom}(X)$. A *relation* R of arity k is any set of instantiations of the form $\{X_a, Y_b, \ldots, Z_c\}$, where $a, b, \ldots, c$ are values from a given universe.

A *constraint* C of arity k is a pair $(\mathrm{vars}(C), \mathrm{rel}(C))$, where $\mathrm{vars}(C)$ is a set of k variables and $\mathrm{rel}(C)$ is a relation of arity k. $I[X]$ denotes the value of X in the instantiation I. $C_{XY...Z}$ denotes a constraint such that $\mathrm{vars}(C) = \{X, Y, \ldots, Z\}$. Given a constraint C, an instantiation I of $\mathrm{vars}(C)$ (or of a superset of $\mathrm{vars}(C)$, considering only the variables in $\mathrm{vars}(C)$), *satisfies* C iff $I \in \mathrm{rel}(C)$. We say that I is *allowed* by C.

A *solution* of a constraint network $\mathcal{N}(\mathcal{X}, \mathcal{D}, \mathcal{C})$ is an instantiation I_S of all variables in $\mathcal{X}$ such that (1.) $\forall X \in \mathcal{X}, I_S[X] \in \mathrm{dom}(X)$ (I_S is *valid*), and (2.) I_S satisfies (is *allowed* by) all the constraints in $\mathcal{C}$.

2.1   Local Consistencies

Definition 1 (Support). *Let C be a constraint and $X \in \mathrm{vars}(C)$. A* **support** *for a value $a \in \mathrm{dom}(X)$ w.r.t. C is an instantiation $I \in \mathrm{rel}(C)$ such that $I[X] = a$.*

Definition 2 (Arc-consistency). *Let C be a constraint and $X \in \mathrm{vars}(C)$. Value $a \in \mathrm{dom}(X)$ is* **arc-consistent** *w.r.t. C iff it has a support in C. C is arc-consistent iff $\forall X \in \mathrm{vars}(C)$, $\mathrm{dom}(X)$ is arc-consistent.*

$\mathcal{N}(\mathcal{X}, \mathcal{D}, \mathcal{C})$ is arc-consistent iff $\forall X \in \mathcal{X}, \forall a \in \mathrm{dom}(X), \forall C \in \mathcal{C}, a$ is arc-consistent w.r.t. C.

Definition 3 (Closure). *Let $\mathcal{N}(\mathcal{X}, \mathcal{D}, \mathcal{C})$ be a constraint network, Φ a local consistency (e.g., AC) and $\mathcal{C}$ a set of constraints $\subseteq \mathcal{C}$. $\Phi(\mathcal{D}, \mathcal{C})$ is the closure of $\mathcal{D}$ for Φ on $\mathcal{C}$, i.e. the set of domains obtained from $\mathcal{D}$ where $\forall X$, all values $a \in \mathrm{dom}(X)$ that are not Φ-consistent w.r.t. a constraint in $\mathcal{C}$ have been removed.*

For GAC and for most consistencies, the closure is unique. In CP systems, a *propagator* is associated with each constraint to enforce GAC or weaker forms of local consistencies. On the other hand, local consistencies stronger than GAC [9,6] require to take into account more than one constraint at a time to be enforced. This fact have made them excluded from most of CP solvers, until now.

2.2   Strong Local Consistencies

This paper focuses on domain filtering consistencies [9], which only prune values from domains and leave the structure of the constraint network unchanged.

Binary Constraint Networks. Firstly, w.r.t. binary constraint networks, as it is mentioned in [6], (i,j)-consistency [11] is a generic concept that captures many local consistencies. A binary constraint network is (i,j)-consistent iff it has non-empty domains and any consistent instantiation of i variables can be extended to a consistent instantiation involving j additional variables. Thus, AC is a $(1,1)$-consistency.

A binary constraint network $\mathcal{N}$ that has non empty domains is :

Path Consistent (PC) iff it is $(2,1)$-consistent.

Path Inverse Consistent (PIC) [12] iff it is $(1,2)$-consistent.

Restricted Path Consistent (RPC) [2] iff it is $(1,1)$-consistent and for all
   values a that have a single consistent extension b to some variable, the pair
   of values (a,b) forms a $(2,1)$-consistent instantiation.

Max-Restricted Path Consistent (Max-RPC) [8] iff it is $(1,1)$-consistent
   and for each value X_a, and each variable $Y \in \mathscr{X} \setminus X$, one consistent extension
   Y_b of X_a is $(2,1)$-consistent (that is, can be extended to any third variable).

Singleton Arc-Consistent (SAC) [3] iff each value is SAC, and a value X_a
   is SAC if the subproblem built by assigning a to X can be made AC (the
   principle is very close to shaving, except that here the whole domains are
   considered).

Non-binary Constraint Networks. Concerning non binary constraint networks, relational arc- and (i,j)-consistencies [10] provide the concepts useful to extend local consistencies defined for binary constraint networks to the nonbinary case. A constraint network $\mathcal{N}$ that has non empty domains is:

Relational AC (relAC) iff any consistent assignment for all but one of the
   variables in a constraint can be extended to the last variable, so as to satisfy
   the constraint.

Relational (i,j)-consistent iff any consistent instantiation for i of the variables in a set of j constraints can be extended to all the variables in the set.

From these notions, new domain filtering consistencies for non-binary constraints inspired by the definitions of RPC, PIC and Max-RPC were proposed in [6]. Moreover, some interesting results were obtained using pairwise consistency. A constraint network $\mathcal{N}$ that has non empty domains is :

Pairwise Consistent (PWC) [14] iff it has no empty relations and any locally consistent instanciation from the relation of a constraint can be consistently extended to any other constraint that intersects with this. One may apply both PWC and GAC.

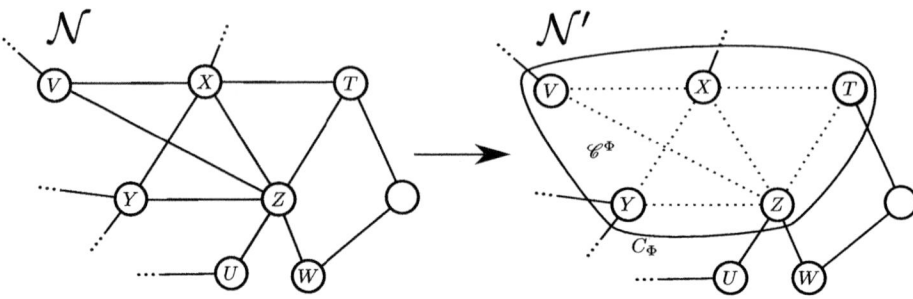

Fig. 1. A strong consistency global constraint C_Φ, used to enforce the strong local consistency on a subset of constraints $\mathscr{C}^\Phi$. $\mathcal{N}'$ is the new network obtained when replacing $\mathscr{C}^\Phi$ by the global constraint.

Pairwise Inverse Consistent (PWIC) [22] iff for each value X_a, there is a support for a w.r.t. all constraints involving X, such that the supports in all constraints that overlap on more variables than X have the same values.

3   A Global Constraint for Domain Filtering Consistencies

This section presents an object-oriented generic scheme for integrating domain filtering consistencies in constraint solvers, and its specialization for Max-RPC. Given a local consistency Φ, the principle is to deal with the subset $\mathscr{C}^\Phi$ of constraints on which Φ should be applied, within a new global constraint C_Φ added to the constraint network. Constraints in $\mathscr{C}^\Phi$ are connected to C_Φ instead of being included into the initial constraint network $\mathcal{N}$ (see Figure 1). In this way, events related to constraints in $\mathscr{C}^\Phi$ are handled in a closed world, independently from the propagation queue of the solver.

3.1   A Generic Scheme

As it is depicted by Figure 2, `AbstractStrongConsistency` is the abstract class that will be concretely specialized for implementing C_Φ, the global constraint that enforces Φ. The constraint network corresponding to $\mathscr{C}^\Phi$ is stored within this global constraint. In this way, we obtain a very versatile framework to implement any consistency algorithm within the event-based solver.

We encapsulate the constraints and variables of the original network in order to rebuild the constraint graph involving only the constraints in $\mathscr{C}^\Phi$, thanks to `SCConstraint` (Strong Consistency Constraint) and `SCVariable` (Strong Consistency Variable) classes. In Figure 1, in $\mathcal{N}'$ all constraints of $\mathscr{C}^\Phi$ are disconnected from the original variables of the solver. Variables of the global constraint are encapsulated in `SCVariables`, and the constraints in `SCConstraints`. In $\mathcal{N}'$, variable Z is connected to the constraints C_{UZ}, C_{WZ} and C_Φ from the point on view of the solver. Within the constraint C_Φ, the `SCVariable` Z is connected to the dotted `SCConstraints` towards the `SCVariables` T, V, X and Y.

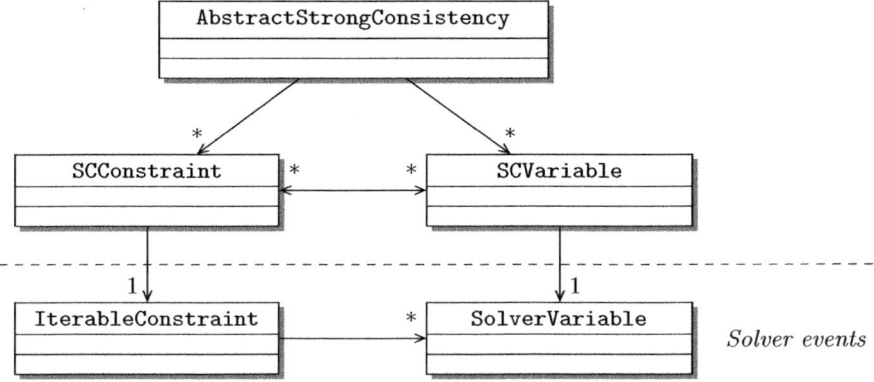

Fig. 2. UML Class diagram [13] of the integration of strong local consistencies into event-based solvers. Arrows describe association relations with cardinalities, either one (1) or many (*).

Note that the original constraints of the problem can be kept in place, so that they can perform their standard pruning task before the stronger consistency is applied. For best efficiency, however, the solver should feature constraint prioritization (see *e.g.*, [26]): propagating the weaker constraints after the strong consistency constraint would be useless.

Mapping the Constraints. We need to identify a lowest common denominator among local consistencies, which will be implemented using the services provided by the constraints of the solver. In Figure 2, this is materialized by the abstract class `IterableConstraint`. Within solvers, and notably event-based solvers, constraints are implemented with *propagators*. While some consistencies such as SAC can be implemented using those propagators, this is not true for most other consistencies. Indeed, the generic concepts that capture those consistencies are (relational) (i, j)-consistencies (see section 2.2). Therefore, they rather rely on the notion of allowed and valid instantiations, and it is required to be able to iterate over and export these, as it is performed to handle logical connectives in [18]. Moreover, algorithms that seek optimal worst-case time complexities memorize which instantiations have already been considered. This usually requires that a given iterator over the instantiations of a constraint always delivers the instantiations in the same order (generally lexicographic), and the ability to start the iteration from any given instantiation.

To give access to and iterate over the supports, the methods `firstSupport` and `nextSupport` are specified in `IterableConstraint`, a subclass of the abstract constraint class of the solver.

Generic iterators. The `firstSupport` and `nextSupport` services are not usually available in most constraint solvers. However, a generic implementation can be

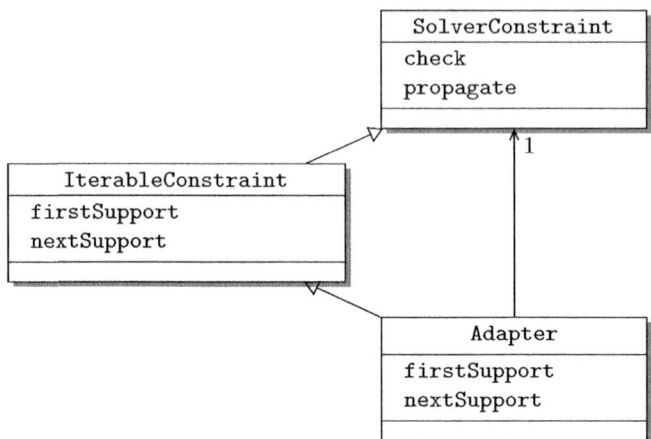

Fig. 3. A generic implementation of support iterator functions, given the constraints provided by a solver. Following the UML specifications, open triangle arrows describe generalization relations.

devised, either by relying on *constraint checkers*[2] (all valid instantiations are checked until an allowed one is found), or by using directly the propagator of the constraint. To perform this, one can simply build a search tree which enumerates the solutions to the CSP composed of the constraint and the variables it involves. These implementations are wrapped in an `Adapter` class that specializes the required `IterableConstraint` superclass, and handles any solver constraint with a constraint checker, as depicted by Figure 3. In this way, no modification is made on the constraints of the solver.

Specialized iterators. For some constraints, more efficient, ad-hoc algorithms for `firstSupport` and `nextSupport` functions can be provided (e.g., for positive table constraints [4]). As `IterableConstraint` specializes `SolverConstraint` (see Figure 3), it is sufficient to specialize `IterableConstraint` for this purpose.

Some strong consistencies such as Path Consistency may be implemented by directly using the propagators of the constraints [16]. Our framework also allows these implementations, since the original propagators of the constraints are still available.

Mapping the Variables. Mapping the variables is simpler, as our framework only requires basic operations on domains, i.e., iterate over values in the current domain and remove values. Class `SCVariable` is used for representing the constraint subnetwork (vars(C_Φ), $\mathcal{D}$, $\mathcal{C}^\Phi$). A link is kept with the solver variable for operation on domains.

---

[2] A constraint checker checks whether a given instantiation is allowed by the constraint or not.

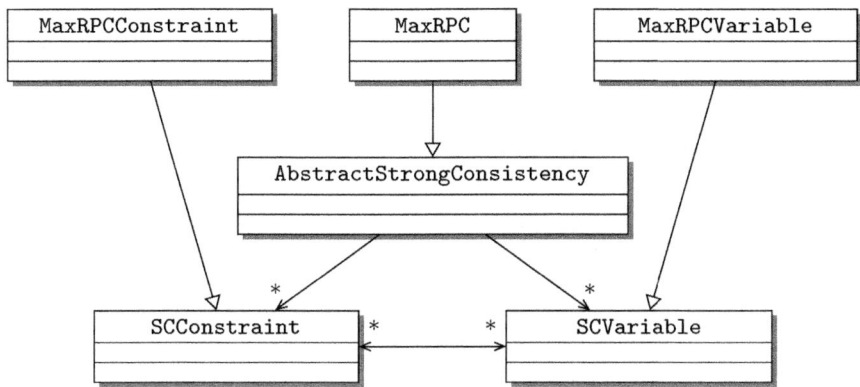

Fig. 4. Diagram of the integration of Max-RPC into event-based solvers

The main feature of `SCVariable` is to "hide" the external constraints from the point of view of the `AbstractStrongConsistency` class implementations. Moreover, it may prove to be very useful to specialize the `SCVariable` class to add data structures required by the strong consistency implementation.

Variable Degree-Based Heuristics. Some popular variable ordering heuristics for binary constraints networks, such as Brelaz, $dom/ddeg$ or $dom/wdeg$, rely on the structure of the constraint graph in order to select the next variable to instantiate. Since constraints in $\mathscr{C}^{\Phi}$ are not connected to the model, they are no longer taken into account by the heuristics of the solver. To overcome this issue, we made the heuristics ask directly for the score of a variable to the `AbstractStrongConsistency` constraints that imply this variable. The global constraint is thus able to compute the corresponding dynamic (weighted) degrees of each variable within their subnetwork $\mathscr{C}^{\Phi}$.

3.2   A Concrete Specialization: Max-RPC

Figure 4 depicts the specialization of our framework to a particular domain filtering consistency for binary networks, Max-RPC [8]. The class `MaxRPC` defines the global constraint that will be used in constraint models. It extends the abstract class `AbstractStrongConsistency` to implement the propagation algorithm of Max-RPC. Moreover, implementing Max-RPC requires to deal with 3-cliques in the constraint graph, to check extensions of a consistent instantiation to any third variable. `SCConstraint` and `SCVariable` classes are specialized to efficiently manipulate 3-cliques.

4   A Coarse Grained Algorithm for Max-RPC

This section presents the implementation of Max-RPC we used in section 5 to experiment our approach.

Algorithm 1. MaxRPC($P = (\mathscr{X}, \mathscr{C}), \mathscr{Y}$)

    $\mathscr{Y}$: the set of variables modified since the last call to MaxRPC

1 $\mathscr{Q} \leftarrow \mathscr{Y}$;
2 **while** $\mathscr{Q} \neq \emptyset$ **do**
3   pick X from $\mathscr{Q}$;
4   **foreach** $Y \in \mathscr{X} \mid \exists C_{XY} \in \mathscr{C}$ **do**
5   **foreach** $v \in \mathrm{dom}(Y)$ **do** **if** revise($C_{XY}, Y_v, \mathrm{true}$) **then** $\mathscr{Q} \leftarrow \mathscr{Q} \cup \{Y\}$;
6   **foreach** $(Y, Z) \in \mathscr{X}^2 \mid \exists (C_{XY}, C_{YZ}, C_{XZ}) \in \mathscr{C}^3$ **do**
7   **foreach** $v \in \mathrm{dom}(Y)$ **do** **if** revisePC(C_{YZ}, Y_v, X) **then** $\mathscr{Q} \leftarrow \mathscr{Q} \cup \{Y\}$;
8   **foreach** $v \in \mathrm{dom}(Z)$ **do** **if** revisePC(C_{YZ}, Z_v, X) **then** $\mathscr{Q} \leftarrow \mathscr{Q} \cup \{Z\}$;

Algorithm 2. revisePC(C_{YZ}, Y_a, X): boolean

    Y: the variable to revise because PC supports in X may have been lost

1 **if** $pcRes[C_{YZ}, Y_a][X] \in \mathrm{dom}(X)$ **then return false** ;
2 $b \leftarrow$ findPCSupport($Y_a, Z_{last[C_{YZ}, Y_a]}, X$) ;
3 **if** $b = \perp$ **then return** revise($C_{YZ}, Y_a, \mathrm{false}$) ;
4 $pcRes[C_{YZ}, Y_a][X] \leftarrow b$; **return false;**

Max-RPC^{rm} [25] is a *coarse-grained* algorithm for Max-RPC. This algorithm exploits backtrack-stable data structures inspired from AC-3^{rm} [17]. *rm* stands for *multidirectional residues*; a residue is a support which has been stored during the execution of the procedure that proves that a given value is AC. During forthcoming calls, this procedure simply checks whether that support is still valid before searching for another support from scratch. The data structures are stable on backtrack (they do not need to be reinitialized nor restored), hence a minimal overhead on the management of data. Despite being theoretically suboptimal in the worst case, Lecoutre & Hemery showed in [17] that AC-3^{rm} behaves better than the optimal algorithm in most cases. In [25], authors demonstrate that using a coarse-grained approach is also especially interesting for the strong local consistency Max-RPC. With g being the maximal number of constraints involving a single variable, c the number of 3-cliques and s the maximal number of 3-cliques related to the same constraint ($s < g < n$ and $e \leq ng/2$), the worst-case time complexity for Max-RPC^{rm} is $O(eg + ed^3 + csd^4)$ and its space complexity is $O(ed + cd)$.

L-Max-RPC^{rm} is a variant of Max-RPC^{rm} that computes a relaxation of Max-RPC with a worst-case time complexity in $O(eg + ed^3 + cd^4)$ and a space complexity in $O(c + ed)$ (that is, a space complexity very close to best AC algorithms). The pruning performed by L-Max-RPC^{rm} is strictly stronger than that of AC.

Algorithms 1 to 4 describe Max-RPC^{rm} and L-Max-RPC^{rm}. In this algorithm, Lines 6-8 of Algorithm 1 and Lines of 5-8 of Algorithm 3 are added to a standard

---

Algorithm 3. revise(C_{XY}, Y_a, *supportIsPC*): boolean

    Y_a: the value of Y to revise against C_{XY} – supports in X may have been lost
    supportIsPC: **false** if one of $pcRes[C_{XY}, Y_a]$ is no longer valid
1   **if** *supportIsPC*$\wedge res[C_{XY}, Y_a] \in$ dom(X) **then return false** ;
2   $b \leftarrow$ firstSupport($C_{XY}, \{Y_a\}$)[X] ;
3   **while** $b \neq \perp$ **do**
4      $PConsistent \leftarrow$ **true** ;
5   **foreach** $Z \in \mathscr{X} \mid (X, Y, Z)$ *form a 3-clique* **do**
6          $c \leftarrow$ findPCSupport(Y_a, X_b, Z) ;
7   **if** $c = \perp$ **then** $PConsistent \leftarrow$ **false** ; **break**;
8          $currentPcRes[Z] \leftarrow c$;
9   **if** $PConsistent$ **then**
10        $res[C_{XY}, Y_a] \leftarrow b$; $res[C_{XY}, X_b] \leftarrow a$;
11        $pcRes[C_{XY}, Y_a] \leftarrow pcRes[C_{XY}, X_b] \leftarrow currentPcRes$;
12   **return false** ;
13      $b \leftarrow$ nextSupport($C_{XY}, \{Y_a\}, \{X_b, Y_a\}$)[$X$] ;
14 remove a from dom(Y) ; **return true** ;

Algorithm 4. findPCSupport(X_a, Y_b, Z): value

---

1   $c_1 \leftarrow$ firstSupport($C_{XZ}, \{X_a\}$)[Z] ;
2   $c_2 \leftarrow$ firstSupport($C_{YZ}, \{Y_b\}$)[Z] ;
3   **while** $c_1 \neq \perp \wedge c_2 \neq \perp \wedge c_1 \neq c_2$ **do**
4   **if** $c_1 < c_2$ **then**
5        $c_1 \leftarrow$ nextSupport($C_{XZ}, \{X_a\}, \{X_a, Z_{c_2-1}\}$)[$Z$] ;
6   **else**
7        $c_2 \leftarrow$ nextSupport($C_{YZ}, \{Y_b\}, \{Y_b, Z_{c_1-1}\}$)[$Z$] ;
8   **if** $c_1 = c_2$ **then return** c_1 ;
9   **return** $\perp$;

AC-3^{rm} algorithm. L-Max-RPC^{rm} removes the memory and time overhead caused by the *pcRes* data structure and the calls to the revisePC function. The principle is to modify Algorithm 1 by removing the **foreach do** loop on Lines 6-8. The revisePC function and *pcRes* data structure are no longer useful and can be removed, together with Lines 8 and 11 of Algorithm 3 (greyed parts in the algorithms). The obtained algorithm achieves an approximation of Max-RPC, which is stronger than AC. It ensures that all the values that were not Max-RPC before the call to L-Max-RPC^{rm} will be filtered. The consistency enforced by L-Max-RPC^{rm} in not monotonous and will depend on the order in which the modified variables are picked from $\mathscr{Q}$, but its filtering power is only slightly weaker than that of Max-RPC on random problems, despite the significant gains in space and time complexities.

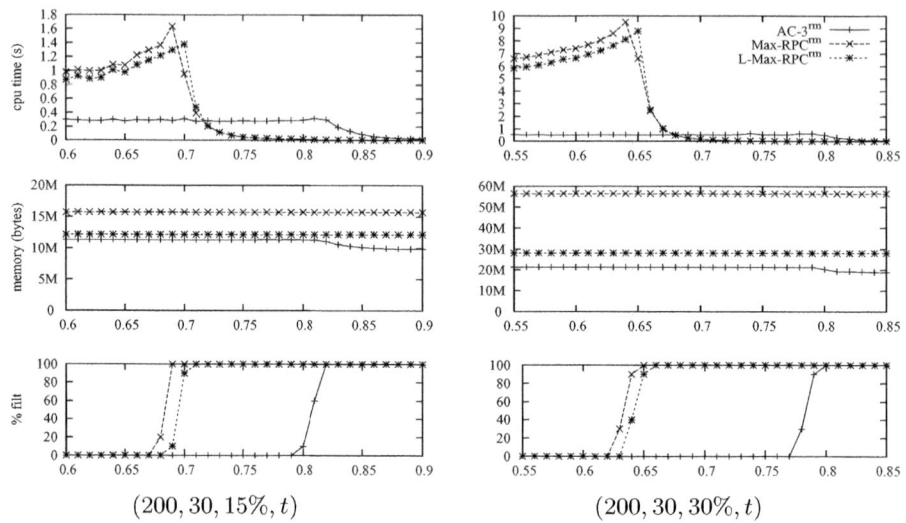

Fig. 5. Initial propagation: CPU time, memory and % of removed values against tightness on homogeneous random problems (200 variables, 30 values, 15/30% density)

5   Experiments

The aim of our experiments is to show the practicability of our approach. We
evaluate (1.) the eventual overload of the integration, and (2.) the interest of
mixing various consistencies, as is made possible thanks to our scheme.

We implemented the diagram of Figure 4 in Choco [15], using the algorithm for Max-RPC described in section 4. In our experiments, Max-RPC^{rm} and
L-Max-RPC^{rm} are compared to Choco's native AC-3^{rm} filtering algorithm.

5.1   Evaluating the Overload

On the figures, each point is the median result over 50 generated binary random
problem of various characteristics. A binary random problem is characterized
by a quadruple (n, d, γ, t) whose elements respectively represent the number of
variables, the number of values, the density[3] of the constraint graph and the
tightness[4] of the constraints.

Single Propagation. Figure 5 compares the time and memory used for the
initial propagation on rather large problems (200 variables, 30 values), as well
as the percentage of removed values. In our experiments, only constraints that
form a 3-clique are mapped to the global constraint. A low density leads to a

---

[3] The density is the proportion of constraints in the graph w.r.t. the maximal number
of possible constraints, i.e. $\gamma = e/\binom{n}{2}$.

[4] The tightness is the proportion of instantiations forbidden by each constraint.

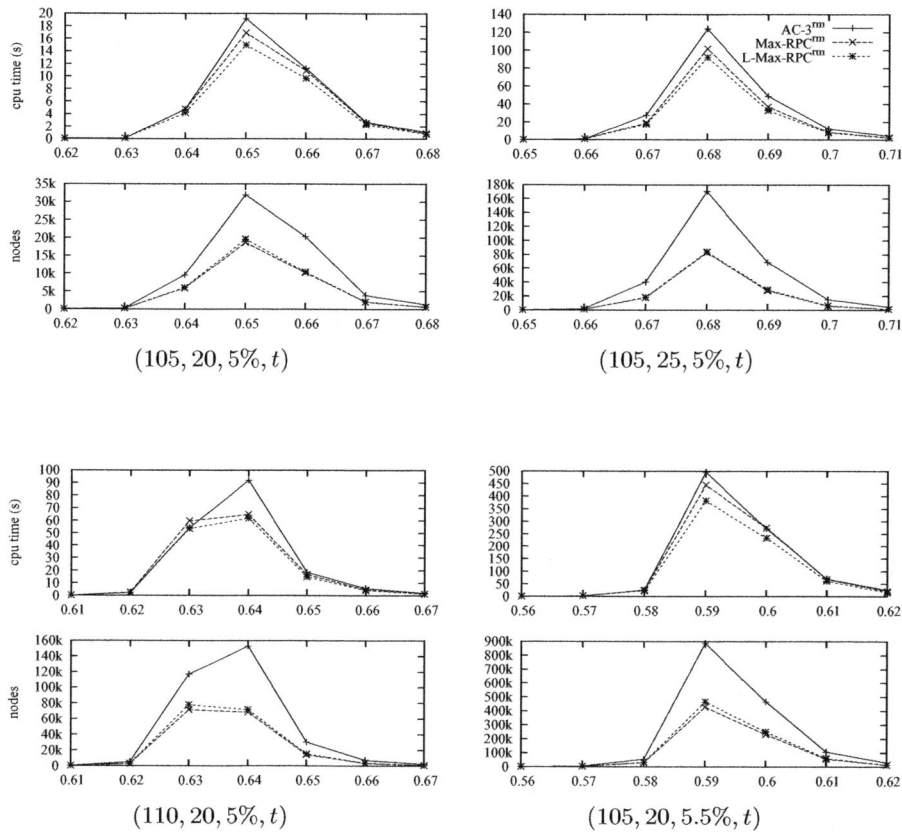

Fig. 6. Full search: cpu time and nodes against tightness on homogeneous random problems (105-110 variables, 20-25 values)

low number of 3-cliques, hence experimental results are coherent with theoretical complexities.

Full Search. Figure 6 depicts experiments with a systematic search algorithm, where the various levels of consistency are maintained throughout search. The variable ordering heuristic is *dom/ddeg* (the process of weighting constraints with *dom/wdeg* is not defined when more than one constraint lead to a domain wipeout). We use the problem $(105, 20, 5\%, t)$ as a reference (top left graphs) and increase successively the number of values (top right), of variables (bottom left) and density (bottom right).

Results in [8,25] showed that maintaining Max-RPC in a dedicated solver was interesting for large and sparse problems, compared with maintaining AC. Our results show that encoding Max-RPC within a global constraint leads to the same conclusions, hence that our scheme has no incidence on computation costs.

Table 1. Mixing two levels of consistency in the same model

		AC-3^{rm}	L-Max-RPC^{rm}	AC-3^{rm}+L-Max-RPC^{rm}
$(35, 17, 44\%, 31\%)$	*cpu (s)*	**6.1**	11.6	non
	nodes	21.4k	8.6k	applicable
$(105, 20, 5\%, 65\%)$	*cpu (s)*	20.0	**16.9**	non
	nodes	38.4 k	19.8 k	applicable
$(35, 17, 44\%, 31\%)$	*cpu (s)*	96.8	103.2	**85.1**
$+(105, 20, 5\%, 65\%)$	*nodes*	200.9k	107.2k	173.4k
$(110, 20, 5\%, 64\%)$	*cpu (s)*	73.0	**54.7**	non
	nodes	126.3k	56.6k	applicable
$(35, 17, 44\%, 31\%)$	*cpu (s)*	408.0	272.6	**259.1**
$+(110, 20, 5\%, 64\%)$	*nodes*	773.0k	272.6k	316.5k

5.2   Mixing Local Consistencies

A new feature provided by our approach is the ability to mix various levels of local consistency for solving a given constraint network, each on some *a priori* disjoint subsets of constraints.[5]

Table 1 shows the effectiveness of the new possibility of mixing two levels of consistency within the same model. The first two rows correspond to the median results over 50 instances of problems $(35, 17, 44\%, 31\%)$ and $(105, 20, 5\%, 65\%)$. The first problem is better resolved by using AC-3^{rm} while the second one shows better results with L-Max-RPC^{rm}.

The third row corresponds to instances where two problems are concatened and linked with a single additional loose constraint. On the last two columns, we maintain AC on the denser part of the model, and L-Max-RPC on the rest. The *dom/ddeg* variable ordering heuristic will lead the search algorithm to solve firstly the denser, satisfiable part of problem, and then thrashes as it proves that the second part of the model is unsatisfiable.

Our results show that mixing the two consistencies entails a faster solving, which emphasizes the interest of our approach. The last two rows present the results with larger problems.

6   Conclusion and Perspectives

This paper presented a generic scheme for adding strong local consistencies to the set of features of constraint solvers. This technique allows a solver to use different levels of consistency for different subsets of constraints in the same model. The soundness of this feature is validated by our experiments. A major

---

[5] Such constraints can share variables.

interest of our schema is that strong consistencies can be applied with different kinds of constraints, including user-defined constraints.

Although our contribution is not restricted to event-based solvers, we underline that an important motivation for providing this scheme was to bridge the gap between strong consistencies and event-based constraint toolkits. Such toolkits put together many scientific contributions of the community. They provide users with advanced APIs that allow to use a catalog of global constraints with powerful filtering algorithms, to implement new constraints, to define specific search strategies, to hybrid CP with other solving techniques such as Local Search (e.g., Comet [24]), or to integrate explanations (e.g., Choco [15]). Our approach adds to this list of features the use of strong consistencies.

Future works include the practical use of our framework with other strong local consistencies, as well as a study of some criteria for decomposing a constraint network, in order to automatize the use of different levels of consistency for different subsets of constraints. This second perspective may allow to link our approach with the heuristics for adapting the level of consistency during the search process [21].

Further, since a given local consistency can be applied only on a subset of constraints, a perspective opened by our work is to identify specific families of constraints for which a given strong consistency can be achieved more efficiently.

References

1. Beldiceanu, N., Carlsson, M., Rampon, J.-X.: Global constraint catalog. Technical Report T2005-08. SICS (2005)
2. Berlandier, P.: Improving domain filtering using restricted path consistency. In: Proceedings of IEEE-CAIA 1995 (1995)
3. Bessière, C., Debruyne, R.: Theoretical analysis of singleton arc consistency and its extensions. Artificial Intelligence 172(1), 29–41 (2008)
4. Bessière, C., Régin, J.-C.: Arc consistency for general constraint networks: preliminary results. In: Proceedings of IJCAI 1997 (1997)
5. Bessière, C., Régin, J.-C.: Enforcing arc consistency on global constraints by solving subproblems on the fly. In: Jaffar, J. (ed.) CP 1999. LNCS, vol. 1713, pp. 103–117. Springer, Heidelberg (1999)
6. Bessière, C., Stergiou, K., Walsh, T.: Domain filtering consistencies for non-binary constraints. Artificial Intelligence 172(6-7), 800–822 (2008)
7. Bessière, C., van Hentenryck, P.: To be or not to be... a global constraint. In: Rossi, F. (ed.) CP 2003. LNCS, vol. 2833, pp. 789–794. Springer, Heidelberg (2003)
8. Debruyne, R., Bessière, C.: From restricted path consistency to max-restricted path consistency. In: Smolka, G. (ed.) CP 1997. LNCS, vol. 1330, pp. 312–326. Springer, Heidelberg (1997)
9. Debruyne, R., Bessière, C.: Domain filtering consistencies. Journal of Artificial Intelligence Research 14, 205–230 (2001)
10. Dechter, R., van Beek, P.: Local and global relational consistency. Theoretical Computer Science 173(1), 283–308 (1997)
11. Freuder, E.C.: A sufficient condition for backtrack-free search. Journal of the ACM 29(1), 24–32 (1982)

12. Freuder, E.C., Elfe, C.D.: Neighborhood inverse consistency preprocessing. In: AAAI/IAAI, vol. 1, pp. 202–208 (1996)
13. Object Management Group. Unified Modeling Language (UML) (2000-2010), http://www.omg.org/spec/UML/
14. Janssen, P., Jegou, P., Nouguier, B., Vilarem, M.C.: A filtering process for general constraint-satisfaction problems: achieving pairwise-consistency using an associated binary representation. In: Proc. of IEEE International Workshop on Tools for Artificial Intelligence, pp. 420–427 (1989)
15. Laburthe, F., Jussien, N., et al.: Choco: An open source Java constraint programming library (2008), http://choco.emn.fr/
16. Lecoutre, C., Cardon, S., Vion, J.: Path Consistency by Dual Consistency. In: Bessière, C. (ed.) CP 2007. LNCS, vol. 4741, pp. 438–452. Springer, Heidelberg (2007)
17. Lecoutre, C., Hemery, F.: A study of residual supports in arc consistency. In: Proceedings of IJCAI 2007, pp. 125–130 (2007)
18. Lhomme, O.: Arc-Consistency Filtering Algorithms for Logical Combinations of Constraints. In: Régin, J.-C., Rueher, M. (eds.) CPAIOR 2004. LNCS, vol. 3011, pp. 209–224. Springer, Heidelberg (2004)
19. Régin, J.-C.: A filtering algorithm for constraints of difference in CSPs. In: Proceedings of AAAI 1994, pp. 362–367 (1994)
20. Régin, J.-C., Petit, T., Bessière, C., Puget, J.-F.: An original constraint based approach for solving over constrained problems. In: Dechter, R. (ed.) CP 2000. LNCS, vol. 1894, pp. 543–548. Springer, Heidelberg (2000)
21. Stergiou, K.: Heuristics for dynamically adapting propagation. In: ECAI, pp. 485–489 (2008)
22. Stergiou, K., Walsh, T.: Inverse consistencies for non-binary constraints. In: Proceedings of ECAI, vol. 6, pp. 153–157 (2006)
23. van Hentenryck, P., Deville, Y., Teng, C.M.: A generic arc-consistency algorithm and its specializations. Artificial Intelligence 57, 291–321 (1992)
24. van Hentenryck, P., Michel, L., See, A., et al.: The Comet Programming Language and System (2001-2007), http://www.comet-online.org
25. Vion, J., Debruyne, R.: Light Algorithms for Maintaining Max-RPC During Search. In: Proceedings of SARA 2009 (2009)
26. Vion, J., Piechowiak, S.: Handling Heterogeneous Constraints in Revision Ordering Heuristics. In: Proc. of the TRICS 2010 Workshop Held in Conjunction with CP 2010 (2010)

Dynamic Constraint Satisfaction Problems: Relations among Search Strategies, Solution Sets and Algorithm Performance⋆

Richard J. Wallace, Diarmuid Grimes, and Eugene C. Freuder

Cork Constraint Computation Centre and Department of Computer Science
University College Cork, Cork, Ireland
{r.wallace,d.grimes,e.freuder}@4c.ucc.ie

Abstract. Previously we presented a new approach to solving dynamic constraint satisfaction problems (DCSPs) based on detection of major bottlenecks in a problem using a weighted-degree method called "random probing". The present work extends this approach and the analysis of the performance of this algorithm. We first show that despite a reduction in search effort, variability in search effort with random probing after problem perturbation is still pronounced, reflected in low correlations between performance measures on the original and perturbed problems. Using an analysis that separates effects based on promise and fail-firstness, we show that such variability is mostly due to variation in promise. Moreover, the stability of fail-firstness is greater when random probiing is used than with non-adaptive heuristics. We then present an enhancement of our original probing procedure, called "random probing with solution guidance", which improves average performance (as well as solution stability). Finally, we present an analysis of the nearest solution in the perturbed problem to the solution found for the original (base) problem. These results show why solution repair methods do poorly when problems are in a critical complexity region, since there may be no solutions similar to the original one in the perturbed problem. They also show that on average probing with solution guidance finds solutions with near-maximal stability under these conditions.

1 Introduction

A "dynamic constraint satisfaction problem", or DCSP, is defined as a sequence of CSPs in which each problem in the sequence is produced from the previous problem by changes such as addition and/or deletion of constraints [VJ05]. Although several strategies have been proposed for handling DCSPs, there is still considerable scope for improvement, in particular, when problems are in the critical complexity region. For these problems, algorithms that attempt to repair the previous solution can be grossly inefficient, especially when they are based on complete search [WGF09].

In recent work we found that for hard CSPs, search performance (amount of effort) can change drastically even after small alterations that do not change the values of the

---

⋆ This work was supported by Science Foundation Ireland under Grant 05/IN/I886. We thank E. Hebrard for contributing the experiment shown in Figure 2.

J. Larrosa and B. O'Sullivan (Eds.): CSCLP 2009, LNAI 6384, pp. 105–121, 2011.

basic problem parameters. At the same time, one feature that is not greatly affected is the set of variables that are the major sources of contention within a problem. It follows that information derived from assessment of these sources of contention should enhance performance even after the problem has been altered. This is what we have found [WGF09]. We also showed that for these problems, a standard DCSP algorithm like Local Changes is 2-3 orders of magnitude worse, depending on the search heuristic.

More specifically, we showed that a heuristic procedure that uses failures obtained during iterated sampling ("random probing") can perform effectively after problem change, *using information obtained before such changes* and thus avoiding the cost of further sampling. The result is a new approach to solving DCSPs based on a robust strategy for ordering variables rather than solution repair or finding robust solutions.

Here, we show that despite this success, predictability of performance across a set of DCSPs of similar character, as reflected in the correlations between original and altered problems, remains fairly low. From an analysis of search performance based on the Policy Framework of [BPW04, BPW05], we find that problem perturbation affects measures of promise to a much greater extent than measures of fail-firstness, which largely explains the anomalous findings.

Based in part on these findings (but also with an eye toward improving solution stability), we developed an enhanced algorithm which uses information in the solution to guide *value* selection (called "random probing with solution guidance"). This allows us to gain the benefits obtained by solution repair techniques such as Local Changes; specifically, search can be strongly limited or avoided entirely when a solution identical to or very similar to the solution to the base problem is in the set of solutions to the perturbed problem.

We also present results of a nearest solution analysis, that is, a comparison of the solution found for the base problem with the closest solution among all solutions to the perturbed problem. This analysis shows that, for problems in the critical complexity region, often there are no similar solutions; this is why solution repair methods perform poorly on average on these problems. On the other hand, random probing with solution guidance can enhance performance whether or not closely similar solutions can be found in the perturbed problem.

The next section gives some background material, including descriptions of the problems used in this study. The third section presents performance data including mean performance and correlations between performance on the base problems and on the perturbed problems. The fourth section presents an analysis in terms of measures of adherence to the promise and fail-first policies. The fifth section describes the random probing with solution guidance procedure. The sixth section presents the nearest solution analysis. The last section gives conclusions.

2   Background Material

2.1   Definitions and Notation

Following [DD88] and [Bes91], we define a dynamic constraint satisfaction problem (DCSP) as a sequence of static CSPs, where each successive CSP is the result of changes in the preceding one. As in earlier work, we consider DCSPs with specific sequence lengths, especially length 1, where "length" is the number of successively altered problems starting from the first alteration to the initial problem.

In our extended notation, $P_{ij}(k)$ refers to the kth member in the sequence for $DCSP_{ij}$, where i is the (arbitrary) number of the initial problem in a set of problems, and j denotes the jth DCSP generated from problem i. However, for DCSPs of length 1 a simpler ij notation is often more perspicuous; in this case Pij is the jth problem (equivalent in this case to the jth DCSP) generated by perturbing base problem i. For mean values (shown in some tables), since i values range over the same set, this notation can be simplified further, to $P\text{-}j$ or Pj.

2.2 Experimental Methods

Because they allow for greater control, many of the present experiments were done with random problems, generated in accordance with Model B [GMP+01]. The base problems had 50 variables, domain size 10, graph density 0.184 and constraint tightness 0.369. Problems with these parameters have 225 constraints in their constraint graphs. Although they are in a critical complexity region, these problems are small enough that they can be readily solved with the algorithms used (together with good heuristics).

For these problems, we restrict our inquiry to the case of addition *and* deletion of k constraints from a base CSP. (Other kinds of change are described in [WGF09].) In this case, the number of constraints remains the same. In addition, changes are carried out so that additions and deletions do not overlap.

DCSP sequences were formed starting with 25 independently generated initial problems. In most experiments, three DCSPs of length 1 were used, starting from the same base problem. Since the effects we observed are so strong, a sample of three was sufficient to show the effects of the particular changes we were interested in.

Search was done with two non-adaptive variable ordering heuristics: maximum forward degree (fd) and the FF2 heuristic of [SG98] $(ff2)$. The latter chooses a variable that maximises the formula $(1 - (1 - p_2^m)^{d_i})^{m_i}$, where m_i is the current domain size of v_i, d_i the future degree of v_i, m is the original domain size, and p_2 is the original average tightness. In addition, adaptive heuristics based on weighted degree were tested, including *dom/wdeg*, *wdeg* [BHLS04], and a version of search using *dom/wdeg* that uses weights at the start of search obtained by "random probing" [GW07]. This latter method involves a number of short 'probes' of the search space where search is run to a fixed cutoff and variable selection is random. Constraint weights are updated in the normal way during probing, but the information is not used until complete search begins. These heuristics were employed in connection with the maintained arc consistency algorithm using AC-3 (MAC-3). The performance measure reported here is search nodes, although constraint checks and runtimes were also recorded.

Experiments on problems with ordered domains involved simplified scheduling problems, used in a recent CSP solver competition[1]. These were "os-taillard-4" problems, derived from the Taillard benchmarks [Tai93], with the time window set to the

---

[1] http://www.cril.univ-artois.fr/~lecoutre/benchmarks/
benchmarks.html

best-known value (os-taillard-4-100, solvable) or to 95% of the best-known value (os-taillard-4-95, insoluble). Each of these sets contained ten problems. For these problems, constraints prevent two operations that form part of the same job or require the same resource from overlapping. Specifically, they are disjunctive relations of the form, $(X_i + dur_i \leq X_j) \bigvee (X_j + dur_j \leq X_i)$, where X_k is the start-time and dur_k the duration of operation k. These problems had 16 variables, the domains were ranges of integers starting from 0, with 100-200 values in a domain, and all variables had the same degree. In this case, the non-adaptive heuristic used was minimum domain/forward degree.

Scheduling problems were perturbed by changing upper bounds of a random sample of domains. In the original problems, domains of the 4-100 problems are all ten units greater than the corresponding 4-95 problems. Perturbed problems were obtained by decreasing four domains of the 4-100 problems by ten units or by increasing six of the domains of the 4-95 problems by ten units. Perturbed problems were selected so that those generated from the 4-100 set remained solvable, while those generated from the 4-95 set remained insoluble.

For solvable problems, the basic demonstrations of DCSP effects were based on a search for one solution. To avoid effects due to vagaries of value selection that might be expected if a single value ordering was used, in these experiments repeated runs were performed on individual problems, with values chosen randomly. (For scheduling problems, value ordering was randomised by choosing either the highest or lowest remaining value in a domain at random.) For random problems, the number of runs per problem was always 100; for scheduling problems with solutions the number was 50. The individual performance datum for each problem is, therefore, mean search nodes over a set of runs.

3   Basic Results

3.1   Non-adaptive Search Heuristics

This section includes some previous results (in some cases extended) to set the stage for the present work. Figure 1 (taken from [WGF09]) shows the extent of variation that can occur after small alterations, in this case addition and deletion of five constraints.

Table 1 shows the grand means for search with fd and $ff2$ across all 75 of the altered CSPs (i.e. the altered problems from the three sets of 25 DCSPs).

Table 1. Search Performance on Altered Problems: Non-Adaptive Heuristics

	fd	$ff2$
5c	2601	3561

Notes. $<50,10,0.184,0.369>$ problems. Single solution search with repeated runs on each problem. "5c" is 5 deletions and additions. Mean search nodes.

Table 2 shows correlations for two sets of experiments (with the two different non-adaptive heuristics), where five constraints were added and deleted. Both here and in

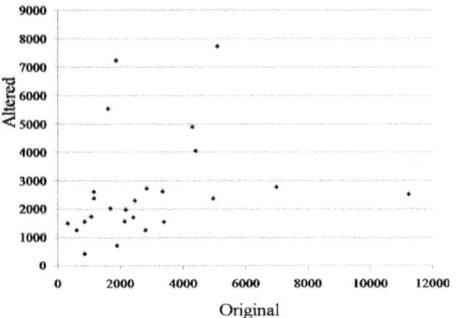

Fig. 1. Scatter plot of search effort (mean nodes over 100 runs) with fd on original versus $P_{i3}(1)$ problems with five constraints added and deleted. (Overall correlation in performance between the 25 original and altered problems is 0.24. Taken from [WGF09].)

Figure 1, vagaries in search effort due to value ordering can be ruled out, since the statistics are based on means of 100 runs per problem with random value selection. In addition, our earlier work ruled out number of solutions as the major factor underlying such variation [WGF09]. These results show that even small changes can have a marked effect on search, and affect the relative difficulty of finding solutions in problems before and after alteration.

Table 2. Correlations with Performance on Original Problems after Alteration: Non-Adaptive Heuristics

	fd			$ff2$		
condit	P1	P2	P3	P1	P2	P3
5c	.49	.83	.24	.34	.54	.31

Notes. $<50,10,0.184,0.369>$ problems. Single solution search with repeated runs on each problem. "5c" is 5 deletions and additions. $Pj = P_{-j}(1)$; thus, P1, P2 and P3 refer to three separate sets of DCSPs of length 1.

Table 3 shows correlations for os-taillard-4 problems for the non-adaptive heuristic, minimum domain over forward degree. Across five sets of perturbed problems, the mean number of search nodes was 257,695 for the 4-95 problems and 391,236 for the 4-100 problems.

Again, there are cases where small changes affect performance to such a degree that the correlations are negligible. That the correlations are often high is not surprising given the nature of the changes to the problems (and the global nature of the correlation coefficient). What is interesting is that even under these conditions changes in performance can occur that are sufficiently marked so that an ensemble measure of similarity can be affected. Moreover, it is possible to obtain marked differences (reflected in negligible correlation coefficients) even for problems without solutions.

Table 3. Correlations with Performance on Original Problems after Alteration for OS-Taillard Problems

problems	P1	P2	d/fd P3	P4	P5
4-95	.03	.98	.99	1.00	.98
4-100	.51	.98	.99	.96	.13

Notes. Problems perturbed by incrementing or decrementing domains. Means for 4-100 problems based on 50 runs per problem with randomised value ordering as described under Methods.

3.2 Adaptive Search Heuristics

A major finding in previous work was that despite their marked effects on search, the changes just described often do not greatly affect the locations of major points of contention (bottleneck variables). Therefore, a heuristic that assesses major sources of contention should perform well, *even when using information from the base problem on a perturbed problem*. Since the constraint weights obtained from random probing distinguish points of high contention [GW07, WG08], the usefulness of this information is not lost in the face of changes such as these. This is shown in Table 4 for the same problems used in Tables 1 and 2. Data are for four methods (the first three of which were used in [WGF09]): (i) *dom/wdeg* with no restarting, (ii) independent random probing for each problem (*rndi*), (iii) a single phase of random probing on the original problems (*rndi-orig*), after which these weights were used with the original and each of its altered problems (on each of the 100 runs with random value ordering), (iv) a strategy in which weights obtained from *dom/wdeg* on the base problem were used at the beginning of search with a perturbed problem. In the third and fourth cases, the new constraints in an altered problem were given an initial weight of 1.

For these problems, random probing improves search performance in comparison with *dom/wdeg*, and there is relatively little fall-off if weights from the base problem are used. However, if weights from *dom/wdeg* are re-used, search performance is distinctly inferior to that found with weights obtained from probing. In fact, it is slightly worse than the original *dom/wdeg* heuristic, which starts search with no information other than degree. This shows the importance of gaining information about contention through random sampling of failures instead of in association with CSP search. It is also consistent with the proposal that random probing provides information about global sources of contention, in (partial) contrast to *dom/wdeg* [GW07].

Table 4. Search Results with Weighted Degree Heuristics: Random Problems

$dom/wdeg$	$rndi$	$rndi\text{-}orig$	$d/wdg\text{-}orig$
1617	1170	1216	1764

Notes. Mean search nodes across all altered problems. First three values from [WGF09].

Table 5 shows correlations for the weighted degree strategies for the same DCSPs as in Tables 1-2. Somewhat surprisingly, these correlations are very similar to the corresponding values found with non-adaptive heuristics. This shows that variability following problem perturbation does not decrease when these strategies are used. The variability is of the same degree when weighted degree ($wdeg$) is used in place of $dom/wdeg$, thus removing effects related to dynamic domain size. At the same time, variability is somewhat less when the original weights from probing are used with the perturbed problems.

A roughly similar situation obtains with the scheduling problems, as shown in Tables 6 and 7.

Table 5. Correlations with Performance on Original Problems after Alteration: Adaptive Heuristics

condit	P1	P2	P3		P1	P2	P3		P1	P2	P3
	$d/wdeg$				$rndi\text{-}d/wdeg$				$rndi\text{-}orig\text{-}d/wdeg$		
5c	.49	.82	.26	|	.58	.76	.38	|	.59	.80	.43
	$wdeg$								$rndi\text{-}orig\text{-}wdeg$		
5c	.40	.82	.29	|				|	.61	.81	.60

Notes. See Table 2.

Table 6. Search Results with Weighted Degree Heuristics: Scheduling Problems

problems	$dom/wdeg$	$rndi$	$rndi\text{-}orig$
taillard-4-95	16,745	4139	5198
taillard-4-100	11,340	7972	5999

Notes. Mean search nodes across all altered problems. Open shop scheduling problems. From [WGF09].

Table 7. Correlations with Performance on Original Problems after Alteration: Adaptive Heuristics

algorithm	P1	P2	P3	P4	P5
	os-taillard-4-95				
d/wdg	.90	.97	.89	.84	.89
$rndi$	.45	.37	.19	.70	.82
$rndi\text{-}orig$	.42	.24	.09	.09	.22
	os-taillard-4-100				
d/wdg	.15	.99	.99	.90	.97
$rndi$	.81	.51	.86	.81	.76
$rndi\text{-}orig$	1.00	.98	.99	.98	.86

Notes. Open shop scheduling problems. P1-P5 refer to five separate sets of DCSPs of length 1.

4   Changes in Promise and Fail-Firstness

The results in the last section leave us with a puzzle: random probing is effective in reducing search effort with perturbed problems, but we still find the same striking variation in performance after small perturbations that we saw with non-adaptive heuristics, reflected in moderate to low correlations between the original and perturbed problems with respect to search effort.

A possible explanation for this discrepancy can be couched in terms of the recently developed Policy Framework of [BPW04, BPW05]. In this framework for backtrack search, search is considered to always be in one of two states: (i) it is on a solution path, i.e. the present partial assignment can be extended to a solution, (ii) a mistake has been made, and search is in an insoluble subtree. In each case, an *ideal* policy for making a decision can be characterised, i.e. a policy that would be optimal if it could actually be realised. In the first case, an optimal policy would maximise the likelihood of remaining on the solution path; in the second, an optimal policy would minimise the size of the refutation (insoluble subtree) needed to prove the incorrectness of the initial wrong assignment. These policies are referred to as the "promise" and "fail-first" policies. Although they cannot be realised in practice, they can be used to characterise good (or bad) performance of variable ordering heuristics. This is because there are measures of adherence to each policy that can be used to assess performance, which are referred to by the same names [BPW04, BPW05, Wal06].

The promise measure is basically a sum of probabilities across all complete search paths. Values can vary between 0 and 1, where a value of 1 means that any value in any domain will lead to a solution. The fail-first measure is the mean "mistake tree" size, where a mistake tree is an insoluble subtree rooted at the first non-viable assignment (i.e. the initial 'mistake'). A larger mean mistake tree size therefore indicates poorer fail-firstness. To avoid artifacts that might arise because of variations in fail-firstness at different search depths, comparisons for fail-firstness were restricted to mistakes made at the same level of search.

It must be emphasized that these measures are not simply assessments of various features of search like mean depth of failure, backtracks at a given level of search, etc. Instead, they are genuine quality-of-search measures, just like total search nodes or run-time. This holds by virtue of their association with two forms of optimal decision making associated with the two conditions of search described above. The rationale for using these policy-based measures is that they give us a more articulated assessment of performance, based on a partition of the states of search into those that can lead to a solution and those that cannot. Put another way, the Policy Framework allows us to define quality-of-search measures specific to each of the two fundamentally distinct types of search-state.

Given this framework, a hypothesis that could explain the improvement in search in spite of continuing variability for individual problems is that the latter is due to variability in promise, i.e. in adherence to the promise policy. Random probing was, in fact, devised as a fail-first strategy [GW07], and previous work has shown that this strategy affects fail-firstness without greatly affecting promise [WG08].

In this work, we used a sampling strategy for assessing adherence to the fail-first policy that is superior to that used in earlier work [BPW05, Wal06]. In earlier work mean mistake tree size was found in connection with an all-solutions search; this meant that sample sizes were not equal either across problems or at different levels of search. Moreover, with this method, there is a confounding factor in that use of a given heuristic above the level of the mistake may affect the efficiency of finding a refutation once a mistake has been made. In the present work, sampling was confined to mistakes at a single level of search, k. Variables and values were chosen at random down to this designated level of search. Below that, search was done with the heuristic being tested. Runs were discarded if a solution was found given the randomly chosen assignments. Otherwise, once search returned to the level of the mistake, the size of the mistake tree (i.e. of the refutation) could be calculated. In addition, for levels > 1, a further criterion was imposed. This is that there had to be a solution based on the original partial assignment for the first k-1 variables assigned. This condition ensures that the mistake at level k is the *first* mistake, and therefore is the true root of the insoluble subtree. If this condition is not met, this means that there is an invalid assignment above level k, and this will affect the average size of the mistake tree as well as introducing unwanted bias into the sampling procedure.

In the present work, the sample size was 100. Thus, for every problem and every level of the first mistake, data were collected for the same number of mistake-trees before obtaining a mean tree size. In addition, seven new sets of DCSPs were tested along with the three original sets.

Table 8. Correlations with Original Problems (Non-adaptive Heuristics)

measure	fd				$ff2$			
	P1	P2	P3	$\overline{x}$(1-10)	P1	P2	P3	$\overline{x}$(1-10)
prom	.22	.74	.34	.46	.50	.43	.16	.41
ff-1	.59	.83	.88	.78	.73	.72	.60	.68
ff-2	.67	.75	.87	.82	.58	.63	.72	.75
ff-3	.82	.80	.77	.81	.70	.83	.80	.78

Notes. $<50,10,0.184,0.369>$ problems. 5 constraints added and deleted. "prom" is promise measure. "ff-k" referes to mistake trees rooted at level k. Mean is for ten sets of DCSPs (the three original sets are also shown individually).

For adaptive heuristics, there is the additional problem that promise and fail-firstness vary in the course of search. To avoid these effects, analysis was restricted to random probing with "frozen" weights, using the weights obtained from the base problems. This means that weights are not updated during search with the perturbed problems. Admittedly, this may elevate the correlations found.

Table 8 (based on the same problems used in Tables 1-2) gives a summary account of these differences in the form of correlations between each successive set of altered problems and the base problem set, for the non-adaptive heuristics, maximum forward degree and FF2. Again, lower correlations reflect changes in magnitude as well as differences in relative magnitude across an entire problem set. These data show that much greater variation occurs in connection with promise than with fail-firstness.

The results in Table 9 show that using heuristics based on contention information still results in low correlations for promise, while correlations for fail-firstness are higher than those found for non-adaptive heuristics.

Table 9. Correlations with Original Problems (Adaptive Heuristic)

measure	P1	P2	P3	$\bar{x}$(1-10)
		rndi-orig-frz		
prom	.82	.66	.22	.36
ff-1	.81	.88	.92	.90
ff-2	.82	.83	.93	.87
ff-3	.79	.87	.86	.86

Notes. See Table 8.

If alterations of this sort have greater effects on promise than fail-firstness, then correlations for search effort before and after perturbation should be also be high when problems have no solutions. (In this case, the only policy in force is the fail-first policy.) This is, in fact, what is found. Table 10 shows results for three sets of perturbed problems; again, correlations are between performance on base problems and each of three sets of perturbed problems. Problems have parameters similar to those used in earlier tables, although the density was increased slightly to reduce the probability of generating problems with solutions.

The weighted-degree heuristics also had consistently high correlations for these insoluble problems. In terms of average nodes over the 75 perturbed problems, there was little fall-off between *rndi-orig* and *rndi* (3812 and 3778 resp.), while both improved over *dom/wdeg* (5268 nodes). These results match those found for the soluble problem set.

Unfortunately, it has not been possible to obtain similar kinds of data for the scheduling problems, using a non-adaptive heuristic like *dom/fwddegree*. This is because of the large number of solutions (making promise calculations difficult) and the large size of the mistake trees for some problems (making fail-firstness calculations difficult).

Table 10. Correlations for Insoluble Problems: Non-Adaptive and Adaptive Heuristics

heuristics	P1	P2	P3
fd	.80	.85	.84
ff2	.90	.91	.93
dom/wdeg	.88	.91	.86
rndi	.84	.86	.89
rndi-orig	.90	.91	.86

Notes. $<$50,10,0.19,0.369$>$ problems. Both base and perturbed problems were insoluble. Perturbations were adding and deleting 5 constraints. "Pj" = $P_{-j}(1)$.

Table 11 gives data on the mean size of insoluble sub-trees for different heuristics for each of the first three levels of search, i.e. when mistakes are made at levels 1, 2, and 3. As one would expect, the size decreases quickly with increasing mistake-depth. In addition, differences in mean size among heuristics correspond to differences in total search nodes found in ordinary search (cf. Tables 1 and 4), although they are higher than values that would be obtained if the first k variables were selected by the same heuristic.

Table 11. Mean Size of Mistake Tree for Each of the First Three Levels of Search

level	fd	$ff2$	$rndi\text{-}orig$
1	2448	3575	1241
2	749	978	394
3	249	312	142

Notes. Means over 275 problems, 100 subtrees per problem at each depth.

5   An Enhanced Probing Procedure

Since variability is already small for fail-firstness, this suggests that further improvements in this direction will be difficult. On the other hand, we do not know of any variable ordering methods that improve promise specifically, since in most cases an ordering that enhances promise also enhances fail-firstness [Wal06]. However, promise can be improved by value orderings as well as by variable orderings (and in fact the concept was only extended to the latter recently [BPW04]).

In this connection, an obvious approach is to piggy-back a solution guidance strategy onto the basic probing procedure. Specifically, we should start testing values in a domain by using the original assignment to that variable. This will allow us to catch those cases where the original solution is still valid, or where there is a solution that is very similar to the original one. We tried this, and in addition we enhanced this strategy with a novel value ordering heuristic, used after the initial value selection. This was to choose as the next candidate assignment the one that was consistent with the highest number of original assignments in the unassigned variables. We call this method "probing with solution guidance". (Note that it is the search that is solution-guided, not the probing. Guiding the probing procedure in this way would, of course, make no sense since it would undermine the sampling strategy.)

With this enhancement mean search effort was reduced significantly, both for *dom/wdeg* and for random probing. For *dom/wdeg* the mean search nodes across 75 perturbed problems (P1-P3) was 954. For $rndi\text{-}orig$ the mean was 701. These are grand means, since the measure of search effort for each problem was the mean of 100 runs in which the base was solved using random value ordering, and the solution obtained was used in the manner described above with the perturbed problem. The reduction in search effort found here was due to the fact that search was greatly abbreviated if the original solution was still valid or if a solution was available that was close to the original with respect to number of common assignments. This also resulted in a smaller mean difference between *dom/wdeg* and $rndi\text{-}orig$.

This new method has the added benefit of improving solution stability (the degree of similarity between the solution to the original problem and the solution to the altered problem). In fact, as shown elsewhere, with this new method there is a small but definite improvement over Local Changes, an algorithm designed to minimise the difference between the old solution and the one found after problem change [VS94].

6   Nearest Solution Analysis

In trying to understand the performance of DCSP algorithms, an important aspect of the problem is the changed solution set after perturbation. Evaluations of this sort may help us understand why solution repair techniques such as Local Changes perform poorly when problems are in the critical complexity region, in contrast to contention-based methods.

Earlier results showed that the number of solutions can change drastically after the kind of change we are considering [WGF09]. However, we also found that number of solutions is only weakly correlated with search effort.

Another aspect of change pertains to the elements in the solution set. If these change appreciably, then expected search effort should also change. Following [VS94], we can assess such change by calculating the Hamming distances between the solution found for the original problem and solutions in the perturbed problem. In particular, given a solution to the base problem, we can determine the Hamming distance of the *nearest* solution (minimal Hamming distance) in the perturbed problem. This can be done using limited discrepancy search (Figure 2) or with branch-and-bound search, where the number of differing assignments serves as the bound. For problems with 50 variables, Hamming distances can range from 0 (meaning the solution found for the base problem is still a solution for the perturbed problem) to 50 (meaning no assignment in the solution to the base problem is part of *any* solution for the perturbed problem).

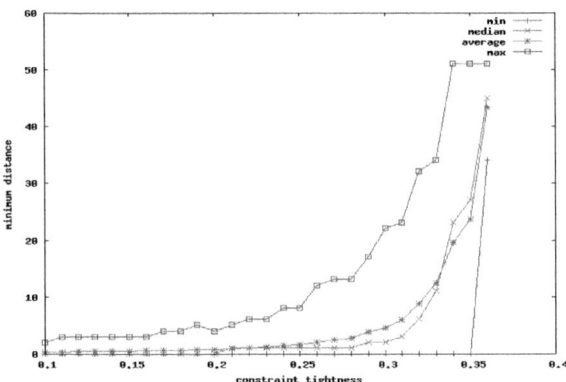

Fig. 2. Distance between the solution to the original problem and the nearest solution in the perturbed problem. Points are based on 100 or 500 randomly generated problems of 50 variables having the given tightness value.

Note that here we are not interested in similarity of the search path, but similarity between solutions. Nor do we need to consider the relation of the changes in assignments to the constraint graph, since, given a solution all constraints are satisfied. Our only concern is the amount of change necessary to employ the new solution rather than the original one, which only involves altering assignments. The amount of such change is, of course, directly measured by the Hamming distance.

Since the problems we have dealt with so far are in the critical complexity region, we first wanted to know how this affected the minimal Hamming distance. In these experiments, problems with the same basic parameters were used, except that constraint tightness was varied. (In addition, only three constraints were added and deleted in each perturbation.) $dom/wdeg$ was used to find a solution to the original problem generated. The most important result is that once one enters the critical complexity region, the minimal Hamming distance rises sharply, and the average approaches the maximum possible (Figure 2).

In subsequent experiments, we used the 5c problems discussed in Section 3. Each of the 25 base problems used previously was solved 100 times using $dom/wdeg$ with random value ordering. For each solution, the minimum Hamming distance was determined for one of the 75 perturbed problems. The average, minimum, maximum and median (minimal) Hamming distance for each problem are shown in Figure 3. For clarity, problems are ordered according to their average minimal Hamming distance over the 100 runs. For the 75 perturbed problems, the average minimal Hamming distance (over 100 runs) varied between 0.4 and 42.7, with a grand mean of 21.2. For 57 of the perturbed problems, there was at least one run where a solution was obtained for the base problem that gave a minimal Hamming distance of 0 when compared with all solutions to the perturbed problem. On the other hand, only 12% of the 7500 runs gave minimal Hamming distances of 0 for this comparison.

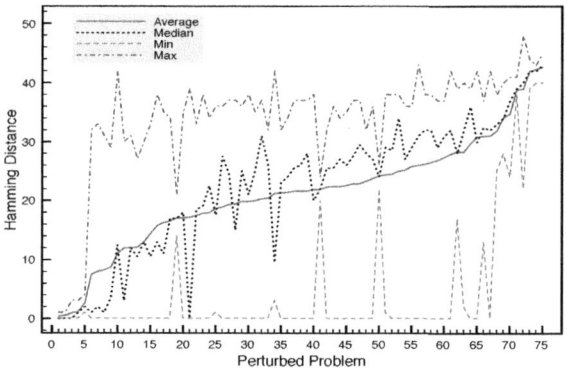

Fig. 3. Hamming distances between solution found for the base problem and closest solution in perturbed problem. Graphs show average minimal distance (solid line) together with the minimum, maximum and median value for each perturbed problem, based on 100 test runs.

Additional insight into these results can be obtained from complete distribution data for individual problems. Figure 4 illustrates the variation in such data using three perturbed problems. For clarity, Hamming distances were ordered from smallest to largest across the hundred runs for each problem. For problem P11-2 (the eleventh problem in the second set of DCSPs), the minimal Hamming distance was always low, indicating that the solution set always contained similar solutions to the one found for the base problem. In contrast, for P6-2 the minimal Hamming distance was always ≥ 40.

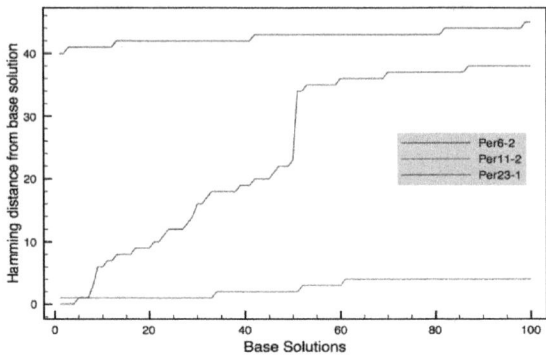

Fig. 4. Hamming distances of nearest solutions based on 100 different solutions to the corresponding base problem, for three different perturbed problems

Most problems fell between these extremes. An example, P23-1, is shown in the figure; here,the minimal Hamming distance ranged from 0 to 38.

These results give a clear indication of why solution reuse methods are often inefficient with hard problems. This is because when problems are hard, often none of the solutions in the perturbed problem are sufficiently close to the original solution for complete methods to gain from solution reuse.

For the three problems whose data are shown in Figure 4, the "weight profiles" (the summed weights of adjacent constraints for each variable) obtained after probing on the original problem and on the perturbed problem were highly correlated (≥ 0.92). This is typical (see [WGF09]), and is, of course, the basis for the strategy of using weights from probes of the base problem with perturbed problems. At the same time, variation in performance after perturbation was found for all three problems, and was not related to the minimal Hamming distances.

On the other hand, a solution reuse procedure like Local Changes showed differences that were related to the minimal Hamming distance. For example, for Local Changes with min-conflicts (look-back) value ordering and $ff2$ for variable selection, nodes explored were 3, 374,495, and 457,750, for problems P11-2, P23-1, and P6-2, respectively. The same ordering was found using fd, although in this case search effort for the latter two problems was one or two orders of magnitude greater.

As might be expected, for probing with solution guidance there is also a clear relation between the nearest solution and performance. This can be seen in Figure 5, which shows the results for individual runs for the three problems, in increasing order of search nodes. (For P11-2, the number of search nodes is always the minimum value of 50, although the minimal Hamming distance is not always 0. This probably reflects the winnowing effects of constraint propagation.)

In this case, these relationships are *not* reflected in a global measure like the correlation coefficient. This is because, as noted above, most problems show a large range for the minimal Hamming distance, depending on the solution found for the base problem. Thus, for Local Changes the correlation between mean minimal Hamming distance and

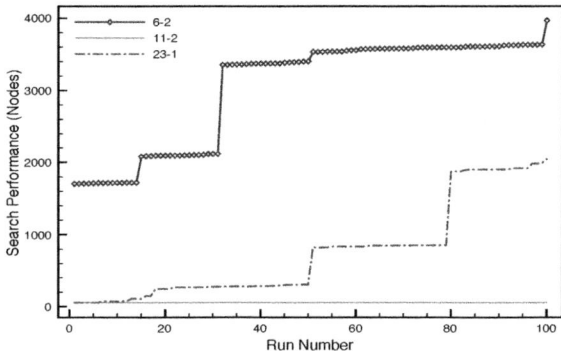

Fig. 5. Performance of probing with solution guidance with base-problem weights across 100 runs for the three perturbed problems in Figure 4

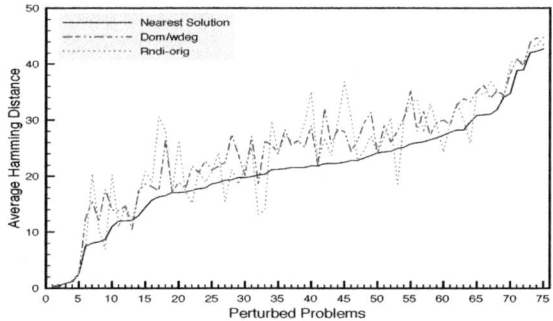

Fig. 6. Solution stability of adaptive heuristics using search with solution guidance, in comparison with average minimal Hamming distance from the earlier tests shown in Figure 3

search performance was 0.24, which is negligible. Interestingly, for $rndi\text{-}orig$ with solution guidance, the correlation was 0.53, still small but appreciably greater than that for Local Changes.

A final analysis shows that for probing with solution guidance, stability is in fact close to optimal. Figure 6 compares assessments of solution stability using the solution-based value ordering heuristic. Assessments were made by calculating the Hamming distance between the solution found for the original problem and that found for the perturbed problem; this was done 100 times for each perturbed problem. These averages are compared with the average minimal Hamming distance found for the same problems in the previous set of tests (cf. Figure 3). In the majority of cases, the means are comparable for each problem. (Naturally, the means based on single solution comparisons are usually higher than the corresponding mean minimal distances.) This indicates that with this method stability is close to the best value possible for these problems.

7 Summary and Conclusions

In earlier work we presented a new approach for solving a range of DCSPs that was much more effective than previous methods. We also presented a rationale for this strategy. However, as shown in the present paper, an apparent paradox arises when we perform the same correlational analysis as we did in earlier work with non-adaptive heuristics. This analysis shows that, in spite of using information about problem features that remain stable in the face of change, we still find marked variation in performance, reflected in modest or low correlations.

In this paper, we consider the hypothesis that, since random probing enhances fail-firstness but not promise, the variability is related to variability in promise. Our results show that this hypothesis is largely correct. Of additional interest is the demonstration that random probing methods give greater predictability in performance with respect to fail-firstness than do non-adaptive heuristics. Here, we would also point out that this analysis demonstrates the usefulness of the Policy Framework for generating *and testing* hypotheses about heuristic performance.

This analysis was also useful in showing us where to look for improvements in performance – by showing us where not to look. In other words, since the issue involved promise rather than fail-firstness, this led us to consider strategies related to the former policy. Guided by this insight, we were able to devise a value ordering strategy which gives a large portion of the benefits of solution repair strategies such as Local Changes (with respect to performance and solution stability), without giving up any of the benefit of our basic contention-based strategy. Thus, mean search effort was reduced because search was avoided in those cases where the old solution was still viable or there was a viable solution that was very close to the original one.

This work also extends the analysis of the characteristics of DCSPs, a hitherto neglected area. We show that for difficult problems the solution set can change dramatically. As a result, it is often the case that there is no solution to the perturbed problem that is similar to the solution found for the original problem. This provides an explanation of why solution repair methods do not fare well for hard problems. This also suggests that the present methods, which rely on assessing the 'deep structure' of the problem, will probably be more robust in general than solution reuse methods.

References

[Bes91]   Bessiére, C.: Arc-consistency in dynamic constraint satisfaction problems. In: Proc. Ninth National Conference on Artificial Intelligence, AAAI 1991, pp. 221–226. AAAI Press, Menlo Park (1991)

[BHLS04]   Boussemart, F., Hemery, F., Lecoutre, C., Sais, L.: Boosting systematic search by weighting constraints. In: Proc. Sixteenth European Conference on Artificial Intelligence, ECAI 2004, pp. 146–150. IOS, Amsterdam (2004)

[BPW04]   Beck, J.C., Prosser, P., Wallace, R.J.: Variable ordering heuristics show promise. In: Wallace, M. (ed.) CP 2004. LNCS, vol. 3258, pp. 711–715. Springer, Heidelberg (2004)

[BPW05]   Beck, J.C., Prosser, P., Wallace, R.J.: Trying again to fail-first. In: Faltings, B., Petcu, A., Fages, F., Rossi, F. (eds.) CSCLP 2004. LNCS (LNAI), vol. 3419, pp. 41–55. Springer, Heidelberg (2005)

[DD88]   Dechter, R., Dechter, A.: Belief maintenance in dynamic constraint networks. In: Proc. Seventh National Conference on Artificial Intelligence, AAAI 1988, pp. 37–42. AAAI Press, Menlo Park (1988)

[GMP+01]   Gent, I.P., MacIntyre, E., Prosser, P., Smith, B.M., Walsh, T.: Random constraint satisfaction: Flaws and structure. Constraints 6, 345–372 (2001)

[GW07]   Grimes, D., Wallace, R.J.: Learning to identify global bottlenecks in constraint satisfaction search. In: Twentieth International FLAIRS Conference, pp. 592–598. AAAI Press, Menlo Park (2007)

[SG98]   Smith, B.M., Grant, S.A.: Trying harder to fail first. In: Proc. Thirteenth European Conference on Artificial Intelligence, ECAI 1998, pp. 249–253. Wiley, Chichester (1998)

[Tai93]   Taillard, E.: Benchmarks for basic scheduling problems. European Journal of Operational Research 64, 278–285 (1993)

[VJ05]   Verfaillie, G., Jussien, N.: Constraint solving in uncertain and dynamic environments: A survey. Constraints 10(3), 253–281 (2005)

[VS94]   Verfaillie, G., Schiex, T.: Solution reuse in dynamic constraint satisfaction problems. In: Twelth National Conference on Artificial Intelligence, AAAI 1994, pp. 307–312. AAAI Press, Menlo Park (1994)

[Wal06]   Wallace, R.J.: Heuristic policy analysis and efficiency assessment in constraint satisfaction search. In: Proc. Eighteenth International Conference on Tools with Artificial Intelligence, ICTAI 2006, pp. 305–312. IEEE Press, Los Alamitos (2006)

[WG08]   Wallace, R.J., Grimes, D.: Experimental studies of variable selection strategies based on constraint weights. Journal of Algorithms: Algorithms in Cognition, Informatics and Logic 63, 114–129 (2008)

[WGF09]   Wallace, R.J., Grimes, D., Freuder, E.C.: Solving dynamic constraint satisfaction problems by identifying stable features. In: Twenty-First International Joint Conference on Artificial Intelligence, IJCAI 2009, pp. 621–627. AAAI/MIT (2009)

Constraint-Based Modeling and Scheduling of Clinical Pathways

Armin Wolf

Fraunhofer FIRST, Kekuléstr. 7, D-12489 Berlin, Germany
armin.wolf@first.fraunhofer.de

Abstract. In this article a constraint-based modeling of clinical pathways, in particular of surgical pathways, is introduced and used for an optimized scheduling of their tasks. The addressed optimization criteria are based on practical experiences in the area of Constraint Programming applications in medical work flow management. Objective functions having empirical evidence for their adequacy in the considered use cases are formally presented. It is shown how they are respected while scheduling clinical pathways.

1 Introduction

Increasing cost in health care and payment on the basis of case-based lump sums (Diagnosis Related Groups – DRG) requires a more efficient, reliable and smooth treatment of patients. One popular approach is the standardization of treatments on the basis of medical work flows – so called *clinical pathways*.

The aim of this paper is the presentation of constraint-based modeling approaches and scheduling techniques for clinical pathways. Clinical pathways are predefined work flows similar to those in traditional (i.e. industrial) scheduling thus suitable for Constraint Programming techniques [3]. In this paper activities in clinical processes, their required resources, their temporal relationships are considered as well as the constraints resulting from the clinical infrastructure and hospital organization. The general structure of clinical pathways and their basics are completed by specialized approaches addressing in particular surgeries. The reason is that surgeries play a central role for the economic situation of a hospital. Well-organized and resource efficient surgical treatment of patients increases work quality and patient satisfaction while decreasing costs and thus the risk of financial loss.

In health care there are particular scheduling approaches for nurse and physician rostering [7,16,17], appointment scheduling [10] surgery scheduling [5] – sometimes with integrated rostering. However, a holistic approach for clinical pathways seems to be missing.

2 Modeling Clinical Pathways

A *clinical pathway* is a standardized medical work flow. It consists of a sequence of *medical activities* like diagnosis, preparation, treatment (e.g. surgery), care,

J. Larrosa and B. O'Sullivan (Eds.): CSCLP 2009, LNAI 6384, pp. 122–138, 2011.

training (e.g. physiotherapy) etc. In general, a medical activity has to be processed within an earliest start time and a latest completion time and within a minimal and maximal duration. Furthermore, a medical activity in general requires several kinds of resources to be performed like nurses, doctors (e.g. surgeons, anesthetists), devices (e.g. MRT), rooms, instruments, consumables etc.

2.1 Tasks

In the context of finite-domain Constraint Programming a medical activity will be represented by at least one *task* formally specified by the following definition:

Definition 1 (Task). *A* task *is a non-preemptive activity with a start time, a duration and an end time. Due to the fact that these times are in general restricted but a-priori undetermined they will be represented by finite-domain variables: Each task t has*

- *a start time variable t.start restricted by t.start $\in S_t$ where S_t is a finite set of non-negative integer values.[1]*
- *a duration variable t.duration restricted by t.duration $\in D_t$ where D_t is a finite set of non-negative integer values.*
- *an end time variable t.end restricted by t.end $\in E_t$ where E_t is a finite set of integer values.*

For each task t the constraint t.start $+ t$.duration $= t$.end has to be satisfied.

*   Furthermore, a task requires some capacity from a resource out of set of resources; thus each task t has*

- *a capacity variable t.capacity restricted by t.capacity $\in C_t$ where C_t is a finite set of non-negative integer values.*
- *a resource identifier variable t.resourceId restricted by t.resourceId $\in R_t$ where R_t is a finite set of resource identifiers (cf. Section 2.2 below).*

In some cases tasks are *optional*, i.e. they are not necessarily performed. This can be modeled either with a possible duration of zero, i.e. duration $\in \{0, p, \ldots, q\}$ where $0 < p \leq q$ holds or with an additional Boolean flag isOptional deciding whether the corresponding task is optional (1 [true]) or mandatory (0 [false]).

2.2 Resources

Constraint-based scheduling knows several kinds of resources. In the context of clinical pathways a task requires in general either

- an *exclusive resource*: it performs at most one activity at the same time, e.g. a surgeon, an operating room, a clinical device etc.
- or an *alternative exclusive resource*: it offers an assortment of exclusive resources, however, one has to be chosen, e.g. similar operating rooms, a collections of the same medical instruments etc.

---

[1] Object-oriented notation is used within this article.

- or a *cumulative resource*: it offers a limited quantity of its capacity, i.e. other tasks can be performed concurrently, e.g. a pool of nurses, consumables, power etc.
- or an *alternative cumulative resource*: it offers an assortment of cumulative resources, however, one has to be chosen, e.g. storages of consumables, pools of nurses etc.

It is assumed that each resource r has a unique integral identifier id_r, i.e. a "number" such that there is a bijection mapping resources to identifiers and vice versa.

In constraint-based scheduling these resources are represented by adequate constraints. If there are cumulative resources the corresponding constraints that has to be satisfied ar defined accordingly:

Definition 2 (Alternative Cumulative Constraint). *Let a set of tasks T and a set of cumulative resources R be given. Each resource $r \in R$ has a unique identifier id_r and an integral capacity $C_r > 0$. Then, the* alternative cumulative constraint *is satisfied for R, T and $T_r = \{t \in T \mid t.\mathsf{resourceId} = id_r\}$ for any $r \in R$ if the following inequality holds:*

$$\forall i \in [\min_{t \in T} S_t, \max_{t \in T} E_t - 1] \, \forall r \in R \, \forall t \in T_r \mid \sum_{t.\mathsf{start} \leq i < t.\mathsf{end}} t.\mathsf{capacity} \leq C_r.$$

It has to be pointed out, that this definition is a special case of the *cumulatives* constraints presented in [4] but generalizes the *cumulative* constraint originally introduced in [1]:

Definition 3 (Cumulative Constraint). *Let a set of tasks T and a cumulative resource r with integral capacity $C_r > 0$ be given such that $R_t = \{r\}$ holds for each task $t \in T$. Then, the* cumulative constraint *is satisfied for T and r if and only if*

$$\forall i \in [\min_{t \in T} S_t, \max_{t \in T} E_t - 1] \, \forall t \in T \mid \sum_{t.\mathsf{start} \leq i < t.\mathsf{end}} t.\mathsf{capacity} \leq C_r$$

holds.

In general, pruning techniques for cumulative constraints, in particular for the *cumulatives* constraint (cf. [4]) support tasks with possibly zero duration and thus optional tasks.

In the special case where consumables have to be represented, cumulative resources can be used for an adequate modeling (cf. [19]). However, there are alternative approaches using specialized algorithms considering consumers as well as producers of consumables to be stored in so-called *reservoirs* or *inventories* with restricted capacities [14,15].

Alternative exclusive resource constraints are special cases of alternative cumulative constraints and exclusive resource constraints are special cases of cumulative constraints. The restriction is that the capacities of all resources is one:

$C_r = 1$ for each considered resource r. Obviously the capacity requirements of all considered tasks must be one, too. Otherwise, the task will neither require the resource (its capacity is zero) nor the constraint will be satisfied because the capacity of the task is greater than one.

However, other definitions are introduced for exclusive constraints in order to highlight the differences with respect to cumulative constraints, in particular there are specialized algorithms to handle these constraints (cf. [3]).

Definition 4 (Alternative Exclusive Resource Constraint). *Let a set of tasks T and a set of exclusive resources R be given. Each task $t \in T$ requires an unary capacity, i.e. t.capacity $= 1$ and each resource $r \in R$ has an unique identifier id_r and an unary capacity $C_r = 1$, too. Then, the* alternative exclusive resource constraint *holds for T and R if*

$$\forall s \in T \, \forall t \in T \setminus \{s\} \, \forall r \in R \mid$$
$$(s.\text{resourceId} = r \wedge t.\text{resourceId} = r) \implies (s.\text{end} \leq t.\text{start} \vee t.\text{end} \leq s.\text{start})$$

is satisfied.

This definition generalizes the definition of exclusive resource constraints:[2]

Definition 5 (Exclusive Resource Constraint). *Let a set of tasks T and an exclusive resource R be given such that t.capacity $= 1$ and $R_t = \{r\}$ holds for each task $t \in T$ and the resource r has an unique identifier id_r and $C_r = 1$ holds. Then, the* exclusive resource constraint *holds for T and r if*

$$\forall s \in T \, \forall t \in T \setminus \{s\} \mid s.\text{end} \leq t.\text{start} \vee t.\text{end} \leq s.\text{start}$$

is satisfied.

In general, pruning techniques for exclusive resource constraints support tasks with possibly zero duration and thus optional tasks. In particular, there are specialized pruning algorithms for optional tasks on exclusive resources [21]. All these algorithms are adopted for alternative exclusive resource constraints (cf. [13,25]).

If a medical activity requires several resources this will be modeled by several tasks – one task for each required (alternative) resource. All these tasks will have identical start times, durations and end times.

Example 1. Within the clinical pathway for a surgery the actual operation op requires an operating room, a surgeon, an anesthetist and some nurses. There are three operating rooms or_1, or_2, or_3, two qualified surgeons su_1, su_2 and four anesthetists $an_1, \ldots, an_4$ available. Two of the five available nurses are required. Obviously the operating room the surgeon and the anesthetist are exclusive resources because they will not treat any other patient while operating. The available nurses are considered as one cumulative resource pool np

---

[2] Exclusive resources are also called "single" resources or "one-machine" resources.

offering a capacity of 5 (nurses). Thus this multi-resource activity *op* will be modeled by four task $op_1, \ldots, op_4$ with op_1.start $\equiv \cdots \equiv op_4$.start, $S_{op_1} \equiv \cdots \equiv S_{op_4}$, op_1.duration $\equiv \cdots \equiv op_4$.duration, $D_{op_1} \equiv \cdots \equiv D_{op_4}$, op_1.end $\equiv \cdots \equiv op_4$.end, and $E_{op_1} \equiv \cdots \equiv E_{op_4}$. Furthermore, op_1.capacity $\in$ $\{1\}$, op_1.resourceId $\in \{id_{or_1}, id_{or_2}, id_{or_3}\}$, op_2.capacity $\in \{1\}$, op_2.resourceId $\in$ $\{id_{su_1}, id_{su_2}\}$, op_3.capacity $\in \{1\}$, op_3.resourceId $\in \{id_{an_1}, \ldots, id_{an_4}\}$, and op_4.capacity $\in \{2\}$, op_4.resourceId $\in \{id_{np}\}$.

In Example 1 all considered resources are independent from each other, thus their identifiers are pairwise different by definition. For a practical implementation of this model in a Constraint Programming system, however, resources are identified by their corresponding constraints and an additional relative index ranging from 0 (or 1) to a maximal index in case of alternative constraints. Consequently, a bijection has to realized that maps identifiers to the according constraints and indexes in case of alternatives.

In clinical practice, often the situation arises that the resources in alternatives are also considered individually or alternative resources share some of their individuals. In both cases, it has to be ensured that all tasks competing for an individual resource in different contexts are related by the according constraints as demonstrated in the following example:

Example 2. Within the clinical pathway for a surgery the actual operation requires that the (a-priori known) leading surgeon is supported by any other (a-priori unknown) assisting surgeon out of a pool of assistants. In general, this pool of assistants consists of surgeons having also leading positions but never both at the same time.

Using the presented modeling approach it is easy to master situations with shared resources:

- use an "all-purpose" alternative cumulative constraint for all cumulative resources and
- use another "all-purpose" alternative exclusive resource constraint for all exclusive resources.

However, it is recommended to use as much as possible "special-purpose" constraints for alternative resources for highest flexibility, i.e. in order to adapt the used pruning algorithms individually or reduce their overall runtime (see below). Thus we suggest to apply the following rules while they are applicable:

- for each individual resource occurring in alternatives use the same identifier and ignore the constraint for the individual resource.[3]
- for any pair of alternatives sharing individual resources use the same identifier for each shared individual and replace the constraints for the pair by one alternative resource constraint covering the individual resources of the pair.

---

[3] It is not really necessary to ignore the redundant constraint.

More formally, let $R_1, \ldots, R_n$ be the possible domains of any resource variables. Then, let $P = P_1 \uplus \ldots \uplus P_m$ be *the partition*[4] of $R_1 \cup \ldots \cup R_n$ such that

- if $R_i \cap P_k \neq \emptyset$ holds for any $i \in \{1, \ldots, n\}$ and any $k \in \{1, \ldots, m\}$ then $R_i \subseteq P_k$ holds, too,
- for each $k \in \{1, \ldots, m\}$ and each non-empty subset $P' \subsetneq P_k$ there is a R_i with $i \in \{1, \ldots, n\}$ such that $R_i \cap P_k \neq \emptyset$ holds, however, $R_i \not\subseteq P'$ holds.

A *Union-Find* algorithm – the classical algorithm was introduced by Tarjan [20] – will compute this partition efficiently. Therefore, we start with the singular sets $\{r_1\}, \ldots, \{r_k\}$ of all considered resources: $\{r_1, \ldots, r_k\} = R_1 \cup \ldots \cup R_n$. Then, for each pair of different resources r_p and r_q the sets containing them are united if there is a set R_i such that $r_p, r_q \subseteq R_i$ holds for $1 \leq i \leq n$ and $1 \leq p < q \leq k$.

Obviously, it holds

$$|T_1|^n + \cdots + |T_m|^n < (|T_1| + \cdots + |T_m|)^n$$

if $T_1, \ldots, T_m$ are the non-empty task sets to be scheduled independently on the resources in $P_1, \ldots, P_m$.[5] Assuming polynomial runtime of the pruning algorithms it means that the overall runtime of the pruning for the "special-purpose" alternative resource constraints is less than the runtime of the pruning for an "all-purpose" alternative resource constraint.

The application of these rules in the situation presented in Example 2 results in the following constraint model:

Example 3 (Continuation of Example 2). Let the clinical pathways for surgeries require an (a-priori known) leading surgeon and sometimes require any other (a-priori unknown) assisting surgeon. Then all potentially leading or assisting surgeons will be considered commonly in an alternative exclusive resource constraint.

2.3   Temporal Relationships

The activities of clinical pathways are in general temporally related. There are absolute/relative and intra-/inter-pathway relationships. In the absolute case tasks can take up to three different roles:

- *Successors*: each successor s of a task t starts after the end of t within a given offset. It holds $t.\text{end} + \text{offset}_{t,s} = s.\text{start}$ where the delay $\text{offset}_{t,s}$ is either an integer value or a finite-domain variable. Choosing the delay appropriately it is possible to model several situations, e.g.
  - $\text{offset}_{t,s} \in [0, \text{MAX\_HORIZON}]$: t must be finished before s will start, which is equivalent to $t.\text{end} \leq s.\text{start}$ if the value MAX_HORIZON is sufficiently large.

---

[4] Such a partition is uniquely defined.
[5] N.B. $T_1, \ldots, T_m$ define a partition as well.

- offset$_{t,s} \in [-5, 5]$: t must be finished at most five time units before/after t will start.

- *Predecessors*: for each predecessor p of a task t it holds that t is a successor of p.
- *Concurrent activities*: for any two concurrent tasks c an d it holds c.start $+$ startOff$_{c,d} = d$.start and/or c.end $+$ endOff$_{c,d} = d$.end where the delays startOff$_{c,d}$ and endOff$_{c,d}$ are either integer values or finite-domain variables. Choosing the delays appropriately it is possible to model several situations:
  - c.start $- 5 = d$.start: c starts five time units later than d
  - c.end $+$ endOff$_{c,d} = d$.end with endOff$_{c,d} \in [0, \text{MAX\_HORIZON}]$: c finishes not-after the end of d.

These relations are complete in the sense of ALLEN [2] because any temporal relationship between the start times and end times of any two a-priori known activities are specifiable. Alternatively, the temporal relationships of start times and end times are representable as a *Simple Temporal Problem* according to [8]:

Definition 6 (Temporal Constraints). *Let a set of start and end time variables E be given. Then the* temporal constraints $T(E)$ *on E are constituted by a set of of inequalities $e - e' \le d$ where d is an integer constant and e, e' are in E.*

The situation becomes more complicated if the (temporal) relations between two medical activities in different clinical pathways depend on the chronological order of the "crucial" tasks of these pathways, e.g. the actual operations in the clinical pathway of two different surgeries. In practice, often the situation arises that a task x_a of a clinical pathway a is related to a task y_b of another clinical pathway b by a constraint $c_{a,b}$, e.g. $c_{a,b} \equiv x_a$.end $+$ offset$_{x_a, y_b} = y_b$.start, if the operating task op_a of pathway a is the direct predecessor of the operating task op_b of pathway b in the commonly used operating room. This means that the conditional constraint between x_a and y_b has to be satisfied only if the operation of a is scheduled directly before the operation of b.

The proposed constraint-based approach to handle this situation adequately uses a sequence dependent setup costs constraint:

Definition 7 (Sequence Dependent Setup Costs). *Let a sequence of tasks $T = t_1, \ldots, t_n$ with $n > 1$ be given. It is assumed that these tasks have to be scheduled in linear order, i.e. it holds*

$$t_i.\text{end} \le t_j.\text{start} \lor t_j.\text{end} \le t_i.\text{start} \quad for \ 1 \le i < j \le n.$$

It is further assumed that for each pair of tasks (t_i, t_j) with $1 \le i, j \le n$ there is an non-negative integer cost value cost$_{i,j}$.

Then, for each task t_j in the sequence T $(1 \le j \le n)$ the finite domain variable setupCost(t_j, T) *defines the* sequence dependent setup cost *of t_j (with respect to T) if this variable is constrained to*

$$
\text{setupCost}(t_j, T) =
\begin{cases}
\text{cost}_{i,j} & \textit{if there is a task } t_i \textit{ with } 1 \leq i \leq n \textit{ such that} \\
& t_i.\text{end} \leq t_j.\text{start } \textit{holds and for each task } t_k \\
& \textit{with } 1 \leq k \leq n, k \neq i, k \neq j \textit{ it holds} \\
& t_k.\text{start} < t_i.\text{end } \textit{or } t_j.\text{start} < t_k.\text{end}, \\
0 & \textit{otherwise.}
\end{cases}
$$

This means that the setup cost of a task t_j is determined by its direct predecessor t_i – if there is any: t_i is scheduled before t_j and there is not any other task scheduled between t_i and t_j.

Now, let $op_1, \ldots, op_n$ be the operations to be performed concurrently in the operating room of op_a and op_b, i.e. $1 \leq a, b \leq n$ holds. Further, let $\text{costValue}_{i,j} = i$ for $1 \leq i, j \leq n$.

Then, it is possible to trigger the constraint $c_{a,b}$ conditionally using *reified constraints*, i.e. logical connections (implication, disjunction etc.) of possibly negated constraints:

$$
(\text{setupCost}(op_b) = a) \implies c_{a,b}.
$$

A rather similar situation is given, if the preparation time for one task in a clinical pathway depends on another task in another pathway. This occurs, e.g. due to cleaning and autoclaving times resulting from a previously performed treatment. An adequate constraint-based model of this situation is usually based on *sequence dependent setup times constraints.* (cf. [9] for a survey and a definition and [24] for a modeling and pruning algorithms).

2.4 Infrastructural and Organizational Relationships

Among the presented treatment-specific relationships there are additional restrictions on the activities of clinical pathways. Some of them are of organizational nature while others are caused by the infrastructure of the medical institution:

- time slots for some activities, sometimes depending on the used resources, e.g. working or operating hours etc.
- (sequence dependent) setup or transfer times, e.g. for autoclaving surgical instruments or for the transportation of the patients from one location to another.
- the location of stations, devices, treatment rooms etc. and the connections between them, e.g. corridors, entrance and exit doors etc.

2.5 Relating Times and Locations

In general, for some tasks there are alternative time slots available in order to process these tasks. However, some of the alternative time slots are only available for some alternative resources and vice versa.

Example 4. On Tuesdays the orthopedic surgeries either has to be performed between 08:00 and 10:30 or between 16:00 and 20:30. However, in the morning there are only the operating rooms numbered 4 and 5 available for orthopedic surgeries and in the afternoon only the rooms with number 1 and 3.

In Constraint Programming it is suitable to model these dependencies between time slots and resources using so called *element* constraints.

Definition 8 (Element Constraint). *Let a sequence of integers values and/or finite-domain variables* $\mathsf{element}_0, \dots, \mathsf{element}_{n-1}$ $(n > 0)$ *be given. Further, let* value *and* index *be two finite domain variables. Then, the* element *constraint*

$$\mathsf{value} = \langle \mathsf{element}_0, \dots, \mathsf{element}_{n-1} \rangle [\mathsf{index}]$$

holds for these entities if value $= \mathsf{element}_{\mathsf{index}}$ *is satisfied.*

Example 5 (Continuation of Example 4). Let $s_1, \dots, s_k$ be the orthopedic surgery tasks to be scheduled next Tuesday according to the spatiotemporal relationships presented in Example 4. The time granularity for scheduling is 5 minutes per time unit. Thus the whole day (0:00 to 24:00) is represented by the interval $[0, 288]$ and e.g. 8 o'clock in morning by 96. It is assumed that each surgery task is additionally defined by

- its variable slot start $\mathsf{slotStart} \in \{96, 192\}$,
- its variable slot end $\mathsf{slotEnd} \in \{126, 246\}$,
- an a.m./p.m. selector variable $\mathsf{ampm} \in \{0, 1\}$,
- a finite-domain variable for the operating rooms available during the morning slot $\mathsf{amOR} \in \{4, 5\}$,
- another finite-domain variable for the operating rooms available during the afternoon slot $\mathsf{pmOR} \in \{1, 3\}$.

Then, for $1 \leq i \leq n$ the constraints

$$s_i.\mathsf{slotStart} = \langle 96, 192 \rangle [s_i.\mathsf{ampm}]$$
$$s_i.\mathsf{slotEnd} = \langle 126, 246 \rangle [s.i.\mathsf{ampm}]$$
$$s_i.\mathsf{resourceId} = \langle s_i.\mathsf{amORs}, s_i.\mathsf{pmORs} \rangle [s_i.\mathsf{ampm}]$$
$$s_i.\mathsf{slotStart} \leq s_i.\mathsf{start}$$
$$s_i.\mathsf{end} \leq s_i.\mathsf{slotEnd}$$

model the spatiotemporal relationships presented in Example 4 adequately. Evidence of this modeling is given by examination of the two cases: $\mathsf{ampm} = 0$ and $\mathsf{ampm} = 1$.

In general, *element* constraints are very appropriate to relate times and/or locations, e.g. to assign the induction and recovery rooms to their according operating room, to assign the transportation times from the hospital wards to the treatment locations etc.

2.6 Room-Specific Relationships

In particular, spatiotemporal relationships in hospitals are sometimes very specific: In one practical application, the situation arises that a newly operated patient treated in one specific operating room has to pass the related induction room in order to end up in the related recovery room. For several reasons (infection risk, privacy etc.) it has to be ensured that there is no other patient in the induction room while moving the newly operated patient from the operating room into the recovery room. – The following constraint-based approach is proposed to consider this "design-specific" situation adequately:

For each surgery i there is a task for induction ind_i and another task for recovery rec_i in its clinical pathway. Obviously, the induction task is before the real surgery task s_i and the recovery task is afterwards:

$$ind_i.\text{start} < ind_i.\text{end} \leq s_i.\text{start} < s_i.\text{end} \leq rec_i.\text{start}.$$

Then, for any two surgeries a and b potentially performed in this special operating room numbered id_{or}, i.e. $R_a \ni id_{or}, R_b \ni id_{or}$, the reified constraint

$$s_a.\text{resourceId} = id_{or} \land s_b.\text{resourceId} = id_{or}$$
$$\implies (rec_b.\text{start} - ind_a.\text{start}) \cdot (rec_a.\text{start} - ind_b.\text{start}) < 0$$

is stated. Alternatively, e.g. if the used Constraint Programming system does not support such reified constraints, the following approach

$$(rec_b.\text{start} - ind_a.\text{start}) \cdot (rec_a.\text{start} - ind_b.\text{start})$$
$$< ((s_a.\text{resourceId} - id_{or})^2 + (s_b.\text{resourceId} - id_{or})^2) \cdot \text{BIG\_INTEGER}.$$

is suitable, too, if the integer value BIG_INTEGER is sufficiently large. In the case that non-linear arithmetics are not supported, another alternative modeling is suggested using the *Kronecker* operator:

Definition 9 (Kronecker Operator). *Let an integer value* val *and a finite-domain variable* var *be given. Then, the* Kronecker *operator is defined by*

$$\delta_{\text{val}}(\text{var}) = \begin{cases} 1 & \textit{if } \text{var} = \text{val}, \\ 0 & \textit{otherwise}. \end{cases}$$

Using the Kronecker operator, the following simplified alternative

$$(rec_b.\text{start} - ind_a.\text{start}) \cdot (rec_a.\text{start} - ind_b.\text{start})$$
$$< (2 - \delta_{id_{or}}(s_a.\text{resourceId}) - \delta_{id_{or}}(s_b.\text{resourceId})) \cdot \text{BIG\_INTEGER}.$$

is also suitable, if the integer value BIG_INTEGER is sufficiently large.

All three approaches models the situation correctly. Evidence is given by the following considerations: If at least one of both surgery tasks is not performed in the considered operating room, then the implication is satisfied and the right-hand-sides of the inequalities become large enough such that they dominate the left-hand-sides of the inequalities. Now, it is assumed that both surgeries are performed in the considered operating room. Then, the *condition* of the implication is satisfied and its *conclusion* is identical with both inequalities, because their left-hand-sides become zero.

Without loss of generality, it is further assumed that surgery a is performed before b.[6] Due to the order of the tasks in the clinical pathways of surgeries it holds ind_a.start $< rec_b$.start. It follows immediately that

$$rec_b\text{.start} - ind_a\text{.start} > 0 \quad \text{and thus} \quad rec_a\text{.start} - ind_b\text{.start} < 0.$$

This means that the recovery of surgery a has already started (in the recovery room after leaving the induction room) before the induction of surgery b starts: rec_a.start $< ind_b$.start.

3   Scheduling Clinical Pathways

The challenge in constraint-based scheduling of clinical pathways is the determination of values for the start, duration, and end variables of all tasks in the pathways such that all all inter- and intra-pathway constraints (see Section 2) are satisfiable, e.g. there are values for the remaining variables such that all constraints are satisfied.

3.1   Optimization

Practical experiences show that optimized scheduling of clinical pathways is rather difficult. Beyond the intrinsic complexity of optimization, the reasons are that the criteria in a clinical context are manifold, informal, and sometimes contradicting each other. There is currently neither a generally accepted objective function nor an according scheduling strategy to be customized via appropriate parameter setting. Different contexts require specialized solutions. Some evidence is given by the following two cases:

Workload Balancing in Surgery Slots. Practical experiences in surgical pathway scheduling have shown that a balancing of the workload within the time slots of the actual surgery tasks is often favored.

The chosen scheduling approach uses the work load factor in percent: workloadFactor $\in [0, 100]$. For a proper consideration of this objective the constraint-based model has to be extended: for each slot s_i with maximal slot duration msd_i (e.g. the difference between the latest slot end and the earliest slot start), with variable slot start slotStart$_i$ and variable slot end slotEnd$_i$ the constraint

$$\text{slotEnd}_i \leq \text{slotStart}_i + \frac{msd_i}{100} \cdot \text{workloadFactor}$$

---

[6] The factors of the product on the left-hand-side are symmetric with respect to the surgeries.

has to be added. More weak criteria like "the surgery tasks should be – if possible – in the chronological order of their requests" following the principle: "first-come, first-served" are difficult to model. Even if penalties for offending this principle are not quantified. In such a case, it is suggested to use an appropriate heuristic search strategy instead of a (strong) branch-and-bound optimization.

This scheduling strategy distributes the surgery tasks over the available slots according to their time and location restrictions (cf. Section 2.5). The algorithm is based on the invariant that the already distributed surgery tasks are constrained to be in linear order in their slots and realizes a depth-first chronological backtracking search:

While there are not yet distributed surgery tasks

- sort the slots according to their work load – lowest load first.
- select the next task according to "first-come, first-served".
- while there is a next slot, try to insert the task into the linear order already established there without causing an inconsistency:
  - start with the last position (after all tasks) first.
  - finish either with success if such position is found or continue with the next slot otherwise.
- if scheduling fails, backtrack to the last recently scheduled task and try another position and/or slot. Otherwise continue with the next task.

This scheduling strategy is completed by a depth-first chronological backtracking labeling strategy. The labeling tries to determine the values of the necessary variables such that all remaining variables will be fixed, too. If the labeling fails, search backtracks to the task distribution until either another distribution of tasks is found or finitely fails because there is no admissible schedule.

A workload balancing distribution of tasks (without linear task orders) can be seen as a preprocessing for a more "local" optimization strategy. Such a strategy on individual operating slots and rooms is addressed in the following:

Optimized Sequencing of Surgeries. In one practical application the challenge is an optimized sequencing of the actual surgery tasks in surgical pathways. In this context the operating rooms and the time slots are already allocated to each surgery task (cf. Section 3.1). Furthermore, there are two operating tables in alternating use: while one table is currently used in the operating room the other is altered for the next surgery. Additionally, all surgeries have priority scores (e.g. high scores for surgeries with a high anesthetic risk) and they have to be performed without break: the next surgery starts directly after the current.

The optimization criteria are manifold:

- minimize the delays of the surgeries according to its scores, i.e. apply the principle: "the higher the score the earlier the surgery",
- reduce the costs for the preparations and alterations of the tables,
- reducing the delays of the surgeries is the primary optimization criterion while the cost reduction has secondary importance.

It is proposed to use a weighted sum to represent all these criteria adequately in an objective function. Therefore, for each surgery s let

- its *priority score* s.score be an integer value,
- its *bedding type* s.type be an integer value determining the sequence dependent preparation/alternation cost for this surgery.

For each bedding type t let $\mathsf{initalCost}_t$ be a non-negative integer value determining the cost for the initial preparation of an operating table for a surgery of type t. For each pair of bedding types s, t let $\mathsf{alterationCost}_{s,t}$ be a non-negative integer value determining the cost for the alteration of an operating table from type s to type t.

Thus, for the sequence of surgeries $S = s_1, \ldots, s_n, (n > 1)$ to be scheduled a permutation $\sigma : \{1, \ldots, n\} \rightarrow \{1, \ldots, n\}$ and a labeling of the start times $s_1.\mathsf{start}, \ldots, s_n.\mathsf{start}$ has to be found such that

$$s_{\sigma(i)}.\mathsf{start} = s_{\sigma(i-1)}.\mathsf{start} + s_{\sigma(i-1)}.\mathsf{duration}$$

is satisfied for $1 \leq i \leq n$ and the sum

$$\mathsf{initalCost}_{s_{\sigma(1)}.\mathsf{type}} + \mathsf{initalCost}_{s_{\sigma(2)}.\mathsf{type}}$$

$$+ \sum_{i=1}^{n-2} \mathsf{alterationCost}_{s_{\sigma(i)}.\mathsf{type}, s_{\sigma(i+2)}.\mathsf{type}} + \sum_{j=1}^{n} \alpha \cdot s_{\sigma(j)}.\mathsf{score} \cdot s_{\sigma(j)}.\mathsf{start}$$

is minimal. – Here, the value of the parameter α adjusts the importance of the delay of surgeries with respect to the costs: the larger the value the higher the importance.

Due to the fact that the order of the surgeries is a-priori unknown, i.e. it has to be determined during an optimized scheduling, appropriate constraints have to be used to model this objective adequately in Constraint Programming. First of all there is a specialized cost constraint required taking the alteration of the operation tables into account:

Definition 10 (Sequence Dependent Alternating Setup Costs). *Let a sequence of tasks $T = t_1, \ldots, t_n$ with $n > 1$ be given. It is assumed that these tasks have to be scheduled in linear order, i.e. it holds*

$$t_i.\mathsf{end} \leq t_j.\mathsf{start} \vee t_j.\mathsf{end} \leq t_i.\mathsf{start} \ \ for \ 1 \leq i < j \leq n.$$

It is further assumed that each task t_i with $1 \leq i \leq n$ has a specific type t.type (an integer value) and there is a non-negative integer cost value $\mathsf{initalCost}_{t.\mathsf{type}}$ and that for each pair of tasks (t_i, t_j) with $1 \leq i, j \leq n$ there is a non-negative integer cost value $\mathsf{alterationCost}_{t_i.\mathsf{type}, t_j.\mathsf{type}}$.

Then, for each task t_j in the sequence T $(1 \leq j \leq n)$ the finite domain variable $\mathsf{alternatingSetupCost}(t_j, T)$ defines the sequence dependent alteration setup cost of t_j (with respect to T) if this variable is constrained to alternatingSetupCost(t_j, T)

$$= \begin{cases} \text{alterationCost}_{t_h.\text{type},t_j.\text{type}} & \textit{if there are two tasks } t_h, t_i \textit{ with } 1 \leq h, i \leq n \\ & \textit{such that } t_h.\text{end} \leq t_i.\text{start} \wedge t_i.\text{end} \leq t_j.\text{start} \\ & \textit{holds and for each task } t_k \textit{ with } 1 \leq k \leq n, \\ & k \neq i, k \neq h, k \neq j \textit{ it holds } t_k.\text{end} \leq t_h.\text{start} \\ & \textit{or } t_j.\text{end} \leq t_k.\text{start}, \\ \text{initialCost}_{t_j.\text{type}} & \textit{otherwise.} \end{cases}$$

This means that the alternating setup cost of a task t_j is determined by the direct predecessor t_h of its direct predecessor t_i – if there are any: t_h is scheduled before t_i, t_i is scheduled before t_j and there is neither a task scheduled between t_h and t_i nor between t_i and t_j.

This constraint is sufficient for a constraint-based modeling of the proposed objective

$$\sum_{i=1}^{n} \text{alternatingSetupCost}(s_i, S) + \alpha \cdot s_i.\text{score} \cdot s_i.\text{start}$$

which is independent of the a-priori unknown permutation and correct by definition of the alternating setup costs constraint and because any permutation of the weighted start times will not change their sum.[7]

3.2   Implementation

The realized scheduling of clinical pathways is based on the presented modeling approaches (cf. Section 2). In particular, schedulers are realized for surgical pathways. The schedulers are implemented in Java combining object-orientation of the host language with Constraint Programming supported by the constraint solver library firstCS [11,23].

The pruning algorithms for the sequence dependent setup costs constraints are based on adjacency matrices. These matrices are used to maintain the transitive closure of the task order relation. For instance, possible predecessors (or pre-predecessors in the case of alternating costs) are used to restrict the domain of a task's cost variable or the maximal gap between two task is used to determine whether a task is a direct predecessor of another task thus determining the cost value of the direct successor task. For sequence dependent setup *times*, however, the implementation of the according constraint is based on the results presented in [6,24].

For the heuristic workload balancing the combined scheduling and labeling presented in Section 3.1 is implemented as described there. For efficiency reasons – this strategy is applied on-line while the surgeries are agreed between the

---

[7] Addition is associative and commutative.

physicians and their patients – this "heuristic optimization" is chosen. The computed schedules are accepted by the physicians and the patients because they are (mostly) conform with their expectations.

The optimized sequencing of surgeries presented in Section 3.1 uses in the first version a weighted task sum constraint – cf. e.g. [12,18] for its definition and pruning algorithms – and a task labeler which labels the start times of the considered tasks after finding a linear task order with depth-first, chronological backtracking search (cf. [12, Section 13.3]). Optimization uses a dichotomic branch-and-bound approach. However, this first approach performs rather bad in some use cases. For improving the runtime of this scheduler a specialized *weighted task sum constraint* [26] is investigated addressing the delay of the surgery start times. The improved pruning algorithms takes into account that the weighted addends are the start times of tasks to be scheduled in linear order yielding better approximations of the sum's lower bound. Furthermore, the labeling is replaced by a simplified search algorithm appropriate for task scheduling without breaks [22]. These modifications performs well even with the chosen dichotomic branch-and-bound optimization of the weighted sum.

Practical applications of the realized scheduling in hospitals have shown that the implementation performs well. In general – even optimal – schedules are computed within a few seconds on a modern PC, however in rare cases the branch-and-bound optimization is interrupted after a individually set time limit yielding sub-optimal but acceptable solutions.

4 Conclusion

There are modelings for clinical pathways presented. The considered modeling aspects reach from intra- to inter-path relationships addressing spatiotemporal and sequence dependent constraints. It is shown how these modelings are applied in practical applications to schedule clinical tasks on medical resources while balancing workload or minimizing costs and delays.

References

1. Aggoun, A., Beldiceanu, N.: Extending CHIP in order to solve complex scheduling and placement problems. Journal of Mathematical and Computer Modelling 17(7), 57–73 (1993)
2. Allen, J.F.: Maintaining knowledge about temporal intervals. Communications of the ACM 26(1), 832–843 (1983)
3. Baptiste, P., le Pape, C., Nuijten, W.: Constraint-Based Scheduling: Applying Constraint Programming to Scheduling Problems. International Series in Operations Research & Management Science, vol. 39. Kluwer Academic Publishers, Dordrecht (2001)
4. Beldiceanu, N., Carlsson, M.: A new multi-resource cumulatives constraint with negative heights. In: Van Hentenryck, P. (ed.) CP 2002. LNCS, vol. 2470, pp. 63–79. Springer, Heidelberg (2002)

5. Belien, J., Demeulemeestera, E.: A branch-and-price approach for integrating nurse and surgery scheduling. European Journal of Operational Research 189(3), 652–668 (2008)
6. Bouquard, J.-L., Lenté, C.: Equivalence to the sequence dependent setup time problem. In: Frank, J., Sabin, M. (eds.) CP 1998 Workshop on Constraint Problem Reformulation (October 30, 1998), http://ti.arc.nasa.gov/ic/people/frank/bouquard.cp98.FINAL.ps (last visited: 2008/08/14)
7. Burke, E.K., De Causmaecker, P., Vanden Berghe, G., Van Landeghem, H.: The state of the art of nurse rostering. The Journal of Scheduling 7(6), 441–499 (2004)
8. Dechter, R., Meiri, I., Pearl, J.: Temporal constraint networks. Artificial Intelligence 49(1-3), 61–95 (1991)
9. Gagné, C., Price, W.L., Gravel, M.: Scheduling a single machine with sequence dependent setup time using ant colony optimization. Technical Report 2001-003, Faculté des Sciences de L'Administration, Université Laval, Québec, Canada (April 2001)
10. Hannebauer, M., Müller, S.: Distributed constraint optimization for medical appointment scheduling. In: Proceedings of the Fifth International Conference on Autonomous Agents, AGENTS 2001, pp. 139–140. ACM, New York (2001)
11. Hoche, M., Müller, H., Schlenker, H., Wolf, A.: firstCS - A Pure Java Constraint Programming Engine. In: Hanus, M., Hofstedt, P., Wolf, A. (eds.) 2nd International Workshop on Multiparadigm Constraint Programming Languages – MultiCPL 2003 (September 29, 2003), http://uebb.cs.tu-berlin.de/MultiCPL03/Proceedings.MultiCPL03.RCoRP03.pdf (last visited: 2009/05/13)
12. Hofstedt, P., Wolf, A.: Einführung in die Constraint-Programmierung, eXamen.press. Springer, Heidelberg (2007) ISBN 978-3-540-23184-4
13. Kuhnert, S.: Efficient edge-finding on unary resources with optional activities. In: Seipel, D., Hanus, M., Wolf, A., Baumeister, J. (eds.) Proceedings of the 17th Conference on Applications of Declarative Programming and Knowledge Management (INAP 2007) and 21st Workshop on (Constraint) Logic Programming (WLP 2007), Institut für Informatik, Am Hubland, 97074 Würzburg, Germany. Technical Report, vol. 434, pp. 35–46. Bayerische Julius-Maximilians-Universität Würzburg (October 2007)
14. Laborie, P.: Algorithms for propagating resource constraints in AI planning and scheduling: Existing approaches and new results. Artifical Intelligence 143, 151–188 (2003)
15. Neumann, K., Schwindt, C.: Project scheduling with inventory constraints. Mathematical Methods of Operations Research 56, 513–533 (2002)
16. Rousseau, L.-M., Pesant, G., Gendreau, M.: A general approach to the physician rostering problem. Annals of Operations Research 115(1), 193–205 (2002)
17. Schlenker, H., Goltz, H.-J., Oestmann, J.-W.: Tame. In: Quaglini, S., Barahona, P., Andreassen, S. (eds.) AIME 2001. LNCS (LNAI), vol. 2101, pp. 395–404. Springer, Heidelberg (2001)
18. Schulte, C., Stuckey, P.J.: When do bounds and domain propagation lead to the same search space. In: Sørgaard, H. (ed.) Third International Conference on Principles and Practice of Declarative Programming, Florence, Italy, pp. 115–126. ACM Press, New York (September 2001)
19. Simonis, H., Cornelissens, T.: Modelling producer/consumer constraints. In: Montanari, U., Rossi, F. (eds.) CP 1995. LNCS, vol. 976, pp. 449–462. Springer, Heidelberg (1995)
20. Trajan, R.E., van Leeuwen, J.: Worst-case analysis of set union algortihms. Journal of the ACM 31(2), 245–281 (1984)

21. Vilím, P., Barták, R., Čepek, O.: Unary resource constraint with optional activities. In: Wallace, M. (ed.) CP 2004. LNCS, vol. 3258, pp. 62–76. Springer, Heidelberg (2004)
22. Wolf, A.: Reduce-to-the-opt – a specialized search algorithm for contiguous task scheduling. In: Apt, K.R., Fages, F., Rossi, F., Szeredi, P., Váncza, J. (eds.) CSCLP 2003. LNCS (LNAI), vol. 3010, pp. 223–232. Springer, Heidelberg (2004)
23. Wolf, A.: Object-oriented constraint programming in Java using the library firstCS. In: Fink, M., Tompits, H., Woltran, S. (eds.) 20th Workshop on Logic Programming, Vienna, Austria, February 22-24. INFSYS Research Report, vol. 1843-06-02, pp. 21–32. Technische Universität Wien (2006)
24. Wolf, A.: Constraint-based task scheduling with sequence dependent setup times, time windows and breaks. In: Im Fokus das Leben, INFORMATIK 2009. Lecture Notes in Informatics (LNI) - Proceedings Series of the Gesellschaft für Informatik (GI), vol. 154, pp. 3205–3219. Gesellschaft für Informatik e.V (2009)
25. Wolf, A., Schlenker, H.: Realizing the alternative resources constraint. In: Seipel, D., Hanus, M., Geske, U., Bartenstein, O. (eds.) INAP/WLP 2004. LNCS (LNAI), vol. 3392, pp. 185–199. Springer, Heidelberg (2005)
26. Wolf, A., Schrader, G.: Linear weighted-task-sum – scheduling prioritised tasks on a single resource. In: Seipel, D., Hanus, M., Wolf, A. (eds.) INAP 2007. LNCS (LNAI), vol. 5437, pp. 21–37. Springer, Heidelberg (2009)

MAC-DBT Revisited

Roie Zivan, Uri Shapen, Moshe Zazone, and Amnon Meisels

Department of Computer Science,
Ben-Gurion University of the Negev,
Beer-Sheva, 84-105, Israel
{zivanr,shapenko,moshezaz,am}@cs.bgu.ac.il

Abstract. Dynamic Backtracking (DBT) is a well known algorithm for solving Constraint Satisfaction Problems. In DBT, variables are allowed to keep their assignment during backjump, if they are compatible with the set of eliminating explanations. A previous study has shown that when DBT is combined with variable ordering heuristics, it performs poorly compared to standard Conflict-directed Backjumping (CBJ) [Bak94]. In later studies, DBT was enhanced with constraint propagation methods. The *MAC-DBT* algorithm was reported by [JDB00] to be the best performing version, improving on both standard DBT and on *FC-DBT* by a large factor.

The present study evaluates the DBT algorithm from a number of aspects. First we show that the advantage of *MAC-DBT* over *FC-DBT* holds only for a static ordering. When dynamic ordering heuristics are used, *FC-DBT* outperforms *MAC-DBT*. Second, we show theoretically that a combined version of DBT that uses both FC and MAC performs equal or less computation at each step than *MAC-DBT*. An empirical result which presents the advantage of the combined version on *MAC-DBT* is also presented. Third, following the study of [Bak94], we present a version of *MAC-DBT* and *FC-DBT* which does not preserve assignments which were jumped over. It uses the *Nogood* mechanism of DBT only to determine which values should be restored to the domains of variables. These versions of *MAC-DBT* and *FC-DBT* outperform all previous versions.

1 Introduction

Conflict Based Backjumping (CBJ) is a technique which is known to improve the search of Constraint Satisfaction Problems (CSPs) by a large factor [Dec03, KvB97, CvB01]. Its efficiency increases when it is combined with forward checking [Pro93]. The down side of CBJ is that when such a backtrack (back-jump) is performed, assignments of variables which were assigned later than the culprit assignment are discarded.

Dynamic Backtracking [Gin93] improves on standard CBJ by preserving assignments of non conflicting variables during back-jumps. In the original form of DBT, the culprit variable which replaces its assignment is moved to be the last among the assigned variables. In other words, the new assignment of the culprit variable must be consistent with all former assignments [Gin93].

Although DBT saves unnecessary assignment attempts and was proposed as an improvement to CBJ, a later study by Baker [Bak94] has revealed a major drawback of DBT. According to Baker, when no specific ordering heuristic is used, DBT performs

J. Larrosa and B. O'Sullivan (Eds.): CSCLP 2009, LNAI 6384, pp. 139–153, 2011.

better than CBJ. However, when ordering heuristics which are known to improve the run-time of CSP search algorithms are used [HE80, BR96, DF02], the performance of DBT is slower than the performance of CBJ. This phenomenon is easy to explain. Whenever the algorithm performs a back-jump it actually takes a variable which was placed according to the heuristic in a high position and moves it to a lower position. Thus, while in CBJ, the variables are ordered according to the specific heuristic, in DBT the order of variables becomes dependent upon the backjumps performed by the algorithm [Bak94].

In order to leave the assignments of non conflicting variables without a change on backjumps, DBT maintains a system of eliminating explanations (*Nogoods*) [Gin93]. As a result, the DBT algorithm maintains dynamic domains for all variables and can potentially benefit from the *Min-Domain* (fail first) heuristic. The present paper demonstrates empirically that this is the best performing version of DBT.

The DBT algorithm was combined with constraints propagation algorithms in order to increase its efficiency. The most successful version reported was *MAC-DBT*. The *MAC-DBT* algorithm uses *support lists* as in the well known $AC4$ algorithm [MH86, BFR95], in order to maintain *Arc Consistency* throughout search. According to [JDB00] *MAC-DBT* outperforms versions of DBT which use a lower level of propagation methods (i.e. Forward Checking). Furthermore, *MAC-DBT* was also reported to outperform former versions of the MAC algorithm [JDB00, BR96].

The present study investigates the DBT algorithm from a number of aspects. First, we show that the advantage of *MAC-DBT* over *FC-DBT* holds only for a static ordering. When dynamic ordering heuristics are used, *FC-DBT* outperforms *MAC-DBT*. Second, we prove theoretically that a combined version of DBT that uses both FC and MAC performs equal or less computation than *MAC-DBT* as presented in [JDB00]. Our empirical results show an advantage of the combined version over *MAC-DBT*. Third, we present a version of *MAC-DBT* which does not preserve assignments which were jumped over (as in standard CBJ). This turns out to be the best performing version of *MAC-DBT*, which we term *MAC-CBJ-NG*. It benefits from the Min-domain heuristic, due to its maintenance of relevant *Nogoods*. Unlike standard *MAC-DBT* it does not harm dynamic ordering by keeping the jumped-over variables assigned. An analogous version of *FC-DBT* is *FC-CBJ-NG*. These versions preserve the properties of the ordering heuristic but in contrast to standard CBJ do not restore removed values whose *Nogoods* are consistent with the partial assignment. These two versions were found to run faster than all previous versions of DBT.

2   Constraint Satisfaction Problems

A *Constraint Satisfaction Problem* (*CSP*) is composed of a set of n variables $V_1, V_2, ..., V_n$. Each variable can be assigned a single value from a discrete finite domain. Constraints or **relations** R are subsets of the Cartesian product of the domains of constrained variables. For a set of constrained variables $\{V_i, V_j, ..., V_m\}$, with domains of values for each variable $\{D_i, D_j, ..., D_m\}$, the constraint is defined as $R \subseteq D_i \times D_j \times ... \times D_m$. A binary constraint R_{ij} between any two variables V_j and V_i is a subset of the Cartesian product of their domains; $R_{ij} \subseteq D_j \times D_i$.

An assignment (or a label) is a pair $\langle var, val \rangle$, where var is a variable and val is a value from var's domain that is assigned to it. A *partial solution* is a consistent set of assignments of values to a set of variables. A **solution** to a *CSP* is a partial solution that includes assignments to all variables [DF02].

3 MAC-DBT

Dynamic Backtracking (DBT) was first introduced in [Gin93]. DBT improves on CBJ by enabeling variables which were jumped over on a backjump to keep their assignments. The DBT algorithm is described next, followed by a description of the later version of [JDB00]which combined MAC with DBT.

3.1 Dynamic Backtracking

The dynamic backtracking (DBT) algorithm is presented following [Bak94]. We assume in our presentation that the reader is familiar with CBJ [Pro93].

Like any backtrack algorithm, DBT attempts to extend a *partial solution*. A partial solution is an ordered set of value assignments to a subset of the CSP variables which is consistent (i.e. violates no constraints). The algorithm starts by initializing an empty partial solution and then attempts to extend this partial solution by adding assigned variables to it. When the partial solution includes assignments to all the variables of the CSP, the search is terminated successfully.

In every step of the algorithm the next variable to be assigned is selected according to the heuristic in use, and the values in its current domain are tested. If a value is in conflict with a previous assignment in the partial solution, it is removed from the current domain and is stored together with its eliminating $Nogood$. Otherwise, it is assigned to the variable and the new assignment is added to the partial solution [Bak94, Gin93].

An order is defined among the assignments in the partial solution. In the simplest form, this order is simply the order in which the assignments were performed (other options will be discussed).

Following [Gin93, Bak94], all $Nogoods$ are of the following form:

$$(v_1 = q_1) \wedge ... \wedge (v_{k-1} = q_{k-1}) \Rightarrow v_k \neq q_k$$

The left hand side serves as the explanation for the invalidity of the assignment on the right hand side. An eliminating $Nogood$ is stored as long as its left hand side is consistent with the current partial solution. When a $Nogood$ becomes invalid, it is discarded and the forbidden value on its right hand side is returned to the current domain of its variable [Gin93].

When a variable's current domain empties, the eliminating $Nogoods$ of all its removed values are resolved and a new $Nogood$ which contains the union of all assignments from all $Nogoods$ is generated. The new $Nogood$ is generated as follows. The right hand side of the generated $Nogood$, includes the assignment which was ordered last in the union of all $Nogoods$ (the culprit assignment). The left hand side is a conjunction of the rest of the assignments in the united set [Gin93, Bak94].

After the new $Nogood$ is generated, all $Nogoods$ of the backtracking variable which include the culprit assignment are removed and the corresponding values are returned to the current domain of the backtracking variable. Notice that this assignment to the backtracking variable cannot possibly be in conflict with any of the assignments which are ordered after the culprit assignment. Otherwise, they would have been included in an eliminating $Nogood$. The culprit variable on the right hand side of the generated $Nogood$ is the next to be considered for an assignment attempt, right after its newly created $Nogood$ is stored. Its new position in the order of the partial solution is after the latest assignment (i.e. it is moved to a lower place in the order than the position it had before).

The variables that were originally assigned after the culprit variable stay assigned. That is in contrast with CBJ (or $FC\text{-}CBJ$ [Pro93]) that discards these assignments.

3.2   DBT with MAC

Enhancing a backtracking algorithm with look ahead methods (FC, AC) provides a significant improvement in performance [Pro93, KvB97, BR96, CvB01] and a solid basis for the ordering heuristics which use the size of domains of unassigned variables [HE80, BR96].

As mentioned above, the DBT algorithm was enhanced with MAC in [JDB00]. In $MAC\text{-}DBT$, after each assignment to a variable v_i, all the values left in the current domain of v_i are entered into a queue (Q) and then an arc consistency method based on $AC4$ is performed. Explanations for the removal of values from the domains of unassigned variables are stored as in standard DBT. For a detailed description of $MAC\text{-}DBT$ the reader is referred to the original paper [JDB00]. Following [JDB00], in our implementation for MAC we used the methodolgy of AC4. That means that we hold a support list for each $[variable, value]$ pair. While this approach is considered less attractive in general for MAC than the most cutting edge algorithms for arc consistency (AC7 for example ensures the space complexity of AC3 while preserving the time complexity of AC4 [BFR95]). The use of support lists is essential in $MAC\text{-}DBT$ for the generation of $Nogood$ eplanations for values which are removed as a result of AC. The explanation for a value removal is explained by the explenations for the removal of its supporters in the corresponding list.

4   Improving MAC-DBT

The contribution of the present paper is centered on improving the $MAC\text{-}DBT$ algorithm in two different directions which are combined into a single algorithm. The first improvement adresses the weakness of the DBT algorithm reported by [Bak94] when ordering heuristics are used. It proposes a version of DBT that does not keep the assignments of "jumped-over" variables. The second improvement adresses the MAC method of [JDB00] and proposes a combined version of FC and MAC to improve its run time.

4.1   CBJ-NG

The $MAC\text{-}DBT$ algorithm of [JDB00] still suffers from the phenomenon reported by [Bak94], that the DBT algorithm abolishes the benefits of the variable ordering heuristic used. Standard CBJ combined with look ahead methods, avoids this problem by discarding the assignment of variables that it jumps-over. This way, the order of variables according to the desired heuristic is preserved. However, in contrast to DBT, in standard CBJ explanations for the removal of values are not stored and therefore on backtracks the entire domain of a variable which was jumped over is restored [Pro93].

The first improvement we propose is to combine DBT and CBJ into *CBJ-NG*. This algorithm benefits from the dynamic ordering of CBJ and from the maintenance of $Nogood$ explanations of DBT. The algorithm uses one of two look ahead methods, FC or MAC, and the resulting algorithms are termed *FC-CBJ-NG* and *MAC-CBJ-NG*, respectively.

During the search process of *FC-CBJ-NG* and *MAC-CBJ-NG* variables are selected to be assigned according to a heursitic. Conflicting values of unassigned variables are filtered out by *forward checking* or by AC. These values are associated with *explicit* nogoods (*Nogoods* which derive explicitly from the problem's initial constraints) that explain their immediate removal as in DBT [Gin93].

When a current domain of an unassigned variable is exhausted, the $Nogood$ explanations of its removed values are resolved and the result is a generation of an *inexplicit* (resolved) $Nogood$. Among the assignments of variables that appear in the generated $Nogood$, the assigned variable that is last ordered in the partial assignment is selected as the target of the backtrack (the *culprit*).

If the culprit is not the last assigned variable, the assignments of variables that were assigned after the culprit are discarded and their assigned values are returned to the current domains of these variables. Eliminating explanations that contain removed or discarded assignment are discarded as well, and the values that were eliminated by them are also returned to their current domains.

In contrast to standard *CBJ*, values of variables, whose assignment was discarded and their $Nogood$ explanation is still consistent with the resulting partial assignment, are *not* returned to their variable's current domain. This forms a solid basis for ordering heuristics that are based on dynamic domain sizes of unassigned variables [HE80, BR96].

4.2   FC-CBJ-NG and MAC-FC-CBJ-NG

The pseudo code of *FC-CBJ-NG* and *MAC-CBJ-NG* is presented in Algorithms 1, 2 and 3. The *italic* lines are code that one needs to perform in order to transform the *FC-CBJ-NG* algorithm into *MAC-FC-CBJ-NG*. The pseudo code of both algorithms is described next.

Procedure **initialize** (line 2) performs initialization of the algorithms' data structures. Most of these operations are basic and technical, therefore they are not described in detail. If *MAC-FC* is performed, the **initialize** procedure computes the support lists for all values [MH86]. In this phase an empty domain generates a report that there is no solution [MH86].

The variables that are handled in lines 3-5 are assumed to be global and accessible by all procedures. $assigned$: is a stack that holds the already assigned variables in a LIFO order of their assignment. $unassigned$: is a pool of the unassigned varibles. The variable $Pseudo$ is intially pushed into the stack, in order to simplify the code. The $unassigned$ pool contains all the variables that participate in the problem. $consistent$ is a boolean variable which indicates whether the problem is consistent.

The loop in lines 6-16 follows the standard form of [Pro93] for CSP search algorithms. Arc consistency is applied before labeling (line 8). The procedure **label**(var) (Algorithm 2) tries to assign one value val from the *current_domain* of the unassigned variable var. If the assignment is successful, consistency remains true, and the assigned variable is entered into the *assigned* stack. Conflicting values are removed from the current domains of future variables. When MAC is used their repspective pairs are inserted into the global queue Q for AC inspection. It is done by the procedure **check_forward**. If the assignment fails (a domain of an unassigned variable empties), state restoration is handled by procedure **undo_reductions**, and val is removed from the current domain of var along with its eliminating $Nogood$. When MAC is performed the pair (var, val) is inserted into the reduction set of the variable that is last assigned. The reduction set of a varaible contains removed values whose eliminating nogoods are consistent with the assignment. These eliminating $Nogoods$ are identified only after the assignment of that variable. The pair (var, val) is also inserted into Q, as the removal of val may cause the problem to violate the arc consistency property. At the end, the current domain of var is inspected for consistency (e.g non-emptiness).

Algorithm 1. MAC-FC/FC-CBJ-NG

```
 1: procedure MAC-FC/FC-CBJ-NG
 2:     initialize()
 3:     unassigned ← variables
 4:     assigned ← pseudoVar
 5:     consistent ← true
 6:     while unassigned.size() > 0 do
 7:         if consistent then
 8:             consistent ← check_AC()
 9:         end if
10:         if consistent then
11:             next_var ← select_next_var(unassigned)
12:             consistent ← label(next_var)
13:         else
14:             consistent ← unlabel()
15:         end if
16:     end while
17:     report solution
18: end procedure
```

Procedure **unlabel** (in Algorithm 2) generally removes the assignment of an assigned variable. A value is removed if the current domain of an unassigned variable empties. The $Nogood$ that explains this removal is resolved (line 2). If the $Nogood$ is empty, the algorithm terminates unsuccessfully (a no solution is reported) (lines 3-6). Otherwise, the right hand side RHS variable, which is called $culprit$, is unassigned. The assignments of variables ordered after the culprit variable are discarded. In the **repeat** loop, these variables are extracted from the $assigned$ stack and a restoration operation is performed by a call to **undo_reductions**. Note that when a variable is extracted its reduction set is unified with the reduction set of the former assigned variable. This

Algorithm 2. Procedures Label and Unlabel

```
 1: procedure LABEL(var)
 2:     select value from var.current_domain
 3:     var.assignment ← value
 4:     consistent ← check_forward(var)
 5:     if not consistent then
 6:         remove var.assignment from var.current_domain
 7:         nogood ← resolve_nogoods(empty_domain_var)
 8:         store(nogood)
 9:         undo_reductions(var)
10:         lastAssigned ← assigned.head( )
11:         add (var,var.assignment) to lastAssigned.reduction
12:         Q ← (var,var.assignment)
13:         if var.current_domain = φ then
14:             empty_domain_var ← var
15:         end if
16:     else
17:         unassigned.remove(var)
18:         assigned.push(var)
19:     end if
20:     return var.current_domain ≠ φ
21: end procedure


 1: procedure UNLABEL
 2:     nogood ← resolve_nogoods(empty_domain_var)
 3:     if nogood = φ then
 4:         report no solution
 5:         stop
 6:     end if
 7:     culprit ← nogood.RHS_variable
 8:     remove var.assignment from var.current_domain
 9:     store(nogood)
10:     repeat
11:         var ← assigned.pop()
12:         undo_reductions(var)
13:         unassigned.add(var)
14:         lastAssigned ← assigned.head( )
15:         add var.reduction to lastAssigned.reduction
16:         if var ≠ culprit then
17:             var.reduction ← φ
18:         else
19:             add (culprit,culprit.assignment) to lastAssigned.reduction
20:             add culprit.reduction to lastAssigned.reduction
21:             Q ← culprit.reduction
22:             culprit.reduction ← φ
23:             if culprit.current_domain = φ then
24:                 empty_domain_var ← culprit
25:             end if
26:             return culprit.current_domain ≠ φ
27:         end if
28:     until var = culprit
29: end procedure
```

Algorithm 3. Procedures FC and check_AC

```
 1: procedure CHECK_FORWARD(var1)
 2:     foreach(var2 ∈ unassigned)
 3:         foreach(val2 ∈ var2.current_domain)
 4:             if not check(var1, var1.assignment, var2, val2) then
 5:                 remove val2 from var2.current_domain
 6:                 nogood ← (var1 = var1.assignment → var2 ≠ val2)
 7:                 store(nogood)
 8:                 Q.add((var2,val2))
 9:                 if var2.current_domain = φ then
10:                     empty_domain_var ← var2
11:                     return false
12:                 end if
13:             end if
14:     return true
15: end procedure


 1: procedure UNDO_REDUCTIONS(culprit)
 2:     foreach(var ∈ unassigned)
 3:         foreach(val ∈ var.domain/var.current_domain)
 4:             nogood ← store.get(var,val)
 5:             if nogood.contains(culprit) then
 6:                 store.remove(nogood)
 7:                 remove((var,val) from culprit.reduction        ▷ if exists..
 8:                 insert val into var.current_domain
 9:             end if
10: end procedure


 1: procedure CHECK_AC                                  ▷ used only in MAC
 2:     while Q ≠ φ do
 3:         (var1,val1) ← Q.extract()
 4:         foreach(var2 ∈ set of variables constrained with var1)
 5:             if var2 ∈ unassigned then
 6:                 foreach(val2 ∈ support(var1,var2,val1))
 7:                     if support(var2,var1,val2) ∩ var1.current_domain = φ then
 8:                         remove val2 from var2.current_domain
 9:                         nogood ← resolve_nogood(support(var2,var1,val2))
10:                         store(var2, nogood)
11:                         lastAssigned ← assigned.head()
12:                         add (var2,val2) to lastAssigned.reduction
13:                         Q.add((var2,val2))
14:                     end if
15:                 if var2.current_domain = φ then
16:                     empty_domain_var ← var2
17:                     return false
18:                 end if
19:             end if
20:     end while
21:     return true
22: end procedure
```

ensures that all reduced values whose eliminating *Nogoods* remain consistent from the time that the culprit was assigned - when the problem was arc consistent - are accumulated. The respective pairs of these values are inserted into Q, and will be examined before the next assignment if the current domain of the culprit variable is consistent.

Procedure **undo_reductions** removes eliminating nogoods that contain the given unassined variable in their explanations. It returns the previously eliminated values to the current domains of the variables they belong to. If MAC is performed these values are removed from the reduction set of the given variable.

Procedure **check_AC** is similar to the main procedure of AC4 [MH86]. It extracts pairs of the form (var, val) from the global queue Q until the queue is emptied. Values of the *current_domain* of unassigned variables, constrained with var, are checked for support (compatible value) in the *current_domain* of var. If there is no support for a value val' of a variable var', then val' is removed and the *Nogood* is resolved. In this case the pair (var', val') is inserted into Q. The pair is also added to the reduction set of the last assigned variable. The procedure stops if it finds an empty *current_domain* or if the Q is emptied. This implies that the problem is arc consistent.

4.3   FC Saves Computation

Let us analyze the computational advantage of performing FC separately, as in Algorithm 1, over the version proposed in [JDB00]. The equivalence of the two methods generates the same removals in unassigned variables. Denote by t, the amount of computations spent by $AC4$ when iterating over the values that are removed from the *current_domain*s of unassigned variables, as a result of direct conflict with the assignment of v_i (all values that would have been removed by FC). Consider the set of all values that belong to *current_domain*s of unassigned variables constrained with v_i. Now, consider a division of this set into twon non intersecting subsets, S - the current set of values that are compatible with the assignment of v_i, and R - the current set of values that are in conflict with that assignment. The FC procedure visits each value in the *current_domain*s of the unassigned variables just once, therefore the time spent by applying FC and than AC is:

$$t + |R| + |S|$$

The method for maintaining AC that is proposed in [JDB00] performs FC by inserting all unassigned values of the *current_domain* of variable i into the Q of removed values. We denote by d, the number of values that were inserted into Q. Each "removed" value triggers a check of support for each $val \in S$. Each $val \in R$ is examined with respect to the number of supporters that are retained in the *current_domain*(v_i). This number is greater or equal to 1. Assuming that the expected number of values checked (lager or equal to the number of supporters) for each $val \in R$, before it is removed from the current domain is k, we conclude that the computations performed are at least:

$$t + d \cdot |S| + k \cdot |R|$$

The cases in which $d = 0$ can be easily checked in advance (no computation needed during search) therefore the values of d and k are always greater or equal to 1. We can easily see that applying FC separately, as proposed in the present paper, generates less or equal computation than the method proposed in [JDB00].

5 Correctness of *MAC-FC-CBJ-NG*

Let us first assume the correctness of the standard DBT algorithm (as proved in [Gin93]) and prove that after the addition of forward checking, of MAC and the elimination of assignments after each backtrack, it is still sound, complete and it terminates.

Soundness is immediate since after each successful assignment the *partial_solution* is consistent. Therefore, when the *partial_solution* includes an assignment for each variable the search is terminated and a consistent solution is reported.

As in the case of standard DBT and *MAC-DBT*, the completeness of *MAC-FC-CBJ-NG* derives from exploring the entire search space except for sub search spaces which were found not to contain a solution. One needs to prove that the sub search spaces which *MAC-FC-CBJ-NG* does not search do not contain solutions. Sub search spaces are pruned by *Nogoods*. It is enough to prove the consistency of the set of *Nogoods* generated by *MAC-FC-CBJ-NG*. In other words, that the assignment of values removed by *Nogoods* never leads to solutions. For standard DBT this is proven in the original paper [Gin93].

The consistency property of *Nogoods* generated by *MAC-FC-CBJ-NG* can be shown as follows. First, observe that during the forward-checking and arc-consistency operations, *Nogoods* are standardly stored as explanations to removed values in the domain of *future variables*. Next, consider the case of a backtrack. It is easy to see that *Nogoods* of the *future variables* are resolved identically to those of standard DBT. Each *Nogood* is either an explicit *Nogood*, which is actually a constraint of the original problem, or a resolved *Nogood* which is a union of explicit *Nogoods*. In both cases any assignment which includes such a *Nogood* cannot be part of a solution. This proves the completeness of *MAC-FC-DBT*.

Last, we need to prove that the algorithm terminates. To this end we need to prove that the algorithm cannot enter an infinite loop. In other words, that a partial assignment cannot be produced by the algorithm more than once. We prove by induction on the number of variables of the CSP, n. For a CSP with a single variable, each of the values is considered exactly once. Assuming correctness of the argument for $CSPs$ with k variables, for all $k < n$, we prove that in the case of a CSP with n variables the argument is still valid. Given a CSP of size n we assign the first variable and prune the inconsistent values of the unassigned variables using FC and MAC. The induced CSP is of size $n - 1$ in which according to the induction assumption the same assignment would not be generated twice. After the search of this induced CSP is completed, the result can be either a solution to the complete CSP, a non solution as a result of the production of an empty *Nogood* or a *Nogood* which includes the assignment of the first variable alone. In the first two cases we are done. In the third case, after the first variable replaced its assignment, it will never assign this value again. Therefore, none of the previous partial assignments can be produced again. This is true for each of the assignments of the first variable.

6 Experimental Evaluation

The common approach in evaluating the performance of *CSP* algorithms is to measure time in logical steps to eliminate implementation and technical parameters from

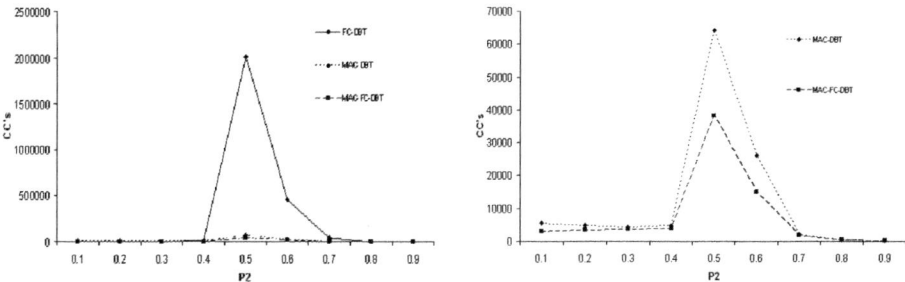

Fig. 1. Constraints checks performed by *FC-DBT*, *MAC-DBT* and *MAC-FC-DBT* on low density CSPs ($p_1 = 0.3$) with static ordering

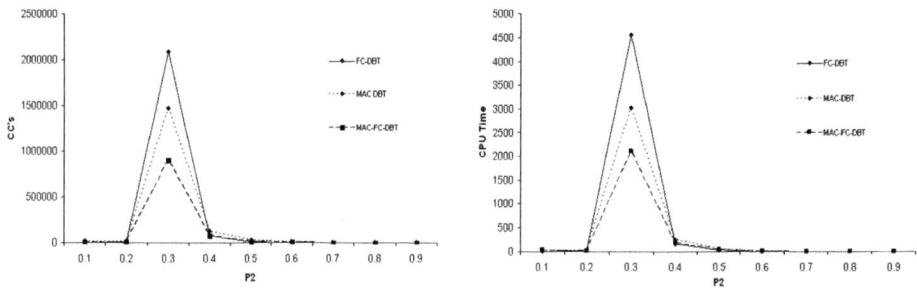

Fig. 2. Same as Figure 1 for high density CSPs ($p_1 = 0.7$)

affecting the results. We present results in both number of constraints checks and in CPU time [Pro96, KvB97].

The experiments were conducted on two problem scenarios: Random *CSPs* and on structured problems that represent a realistic scenario - Meeting Scheduling Problems [GW99].

Random *CSPs* are parametrized by n variables, k values in each domain, a constraints density of p_1 and a tightness p_2 which are commonly used in experimental evaluations of CSP algorithms [Smi96]. Two sets of experiments were performed on random problems. Both were conducted on *CSPs* with 20 variables ($n = 20$) and 10 values in the domain of each variable ($k = 10$). Two values of constraints density were used, $p_1 = 0.3$ and $p_1 = 0.7$. The tightness value p_2, was varied between 0.1 and 0.9, in order to cover all ranges of problem difficulty. For each of the pairs of fixed density and tightness (p_1, p_2), 50 different random problems were solved by each algorithm and the results presented are an average of these 50 runs.

In the first set of experiments, DBT with three different lookahead methods was compared, Forward Check (*FC-DBT*), MAC (*MAC-DBT*) and a combined lookahead version that use both FC and MAC (*MAC-FC-DBT*).

The left hand side (LHS) of Figure 1 presents the number of constraints checks performed by the three versions of the algorithm with static ordering on low density CSPs ($p_1 = 0.3$). *MAC-DBT* outperforms *FC-DBT*, as reported by [JDB00]. The algorithm

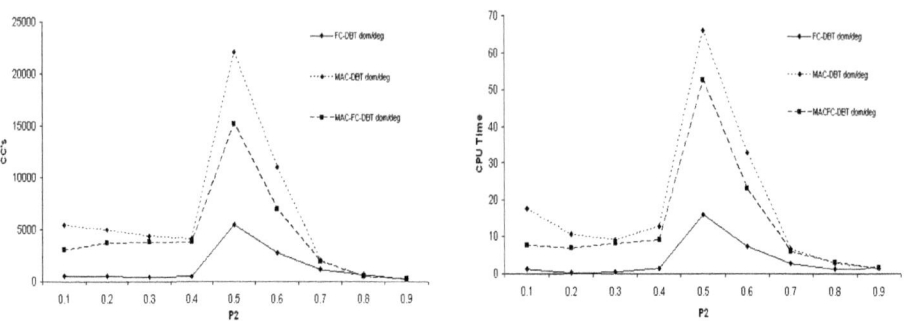

Fig. 3. Constraints checks and CPU-time performed by *FC-DBT*, *MAC-DBT* and *MAC-FC-DBT* with *dynamic* (min-domain / degree) ordering ($p_1 = 0.3$)

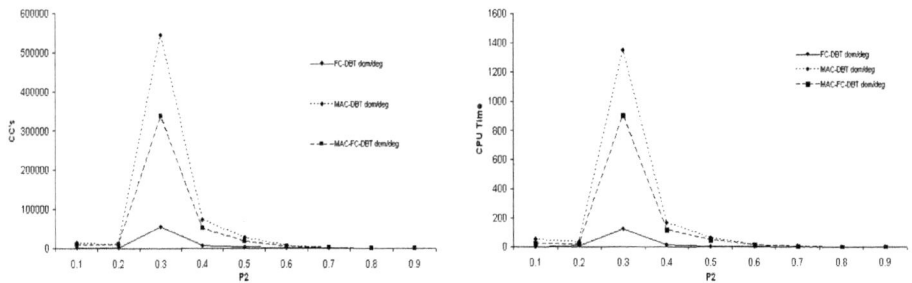

Fig. 4. Same as Figure 5 for high density CSPs ($p_1 = 0.7$)

proposed in the present paper (that performs *FC* seperately of *AC4*) *MAC-FC-DBT* outperfroms both *MAC-DBT* and *FC-DBT*. A closer look at the difference between *MAC-FC-DBT* and *MAC-DBT* is presented on the right hand side of the (RHS)figure. Note that in this experiment all algorithms perform *DBT* and not *CBJ-NG*. They all preserve the jumped over assignments on a backtrack as in [Gin93, Bak94, JDB00].

The LHS of Figure 2 presents similar results for high density CSPs ($p_1 = 0.7$). In this case the difference between *FC-DBT* and *MAC-DBT* is much smaller. The RHS of Figure 2 presents the same results in CPU time and is presented in order to show the similarity between these two measures.

Figure 3 presents a comparison between the same versions of the algorithm when using the min-domain/degree heuristic [BR96]. The results in constraints checks (LHS) and in CPU time (RHS) show clearly the advantage of *FC-DBT* on both versions of *MAC-DBT* when ordering heuristics are used. Similar results for high density CSPs are presented in Figure 4. The difference in favor of *FC* is higher on dense CSPs.

In the second set of experiments *FC-DBT* and *MAC-FC-DBT* are compared with the well known *FC-CBJ* algorithm and with the two proposed versions which perform backjumping as in CBJ but store $Nogoods$ as in DBT (see Section 3.1), our proposed algorithms *MAC-FC-CBJ-NG* and *FC-CBJ-NG* (Algorithms 1, 2 and 3). All algorithms use the min-domain/degree heuristic.

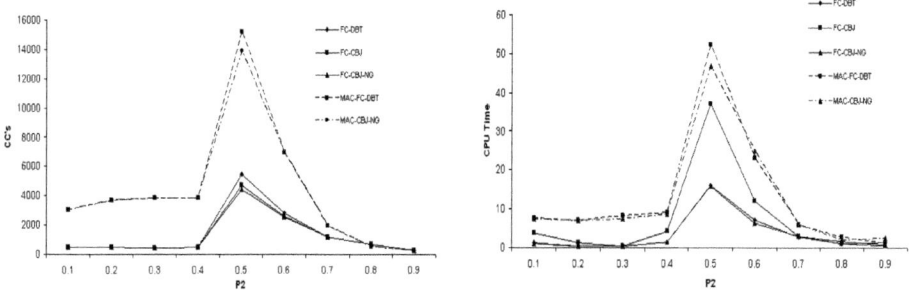

Fig. 5. Constraints checks and CPU-time for low density CSPs ($p_1 = 0.3$)

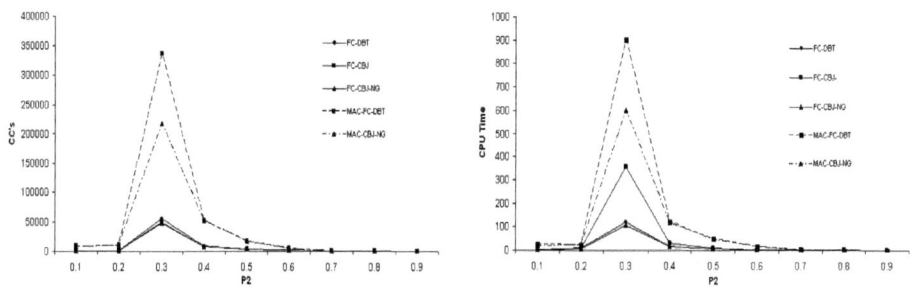

Fig. 6. Same as Figure 3 for high density CSPs ($p_1 = 0.7$)

Figure 5 presents a comparison between different versions of the algorithm on low density random $CSPs$. The results show a large difference between the FC algorithms and the MAC algorithms. However, there are small differences between the different versions. Similar results were obtained for high density random $CSPs$ (Figure 6).

In the third set of experiments, the algorithms are compared when solving structured problems, Meeting Scheduling [GW99]. In this special class of problems, variables represent meetings between agents. Arrival constraints exist between meetings of the same agent. The tightness of the problem grows with the number of meetings per agent [GW99].

Figure 7 presents the performance of the algorithms, solving random meeting scheduling problems with 40 meetings (variables), domain size of 12 time slots, 18 agents and arrival constraints which were randomly selected between 2 to 6 [GW99]. The results in constraints checks (LHS) and in CPU-time are again very similar. On structured problems the differences between the different versions of MAC-DBT are much smaller than the differences between the different versions of FC. Still, the best performing algorithm is the FC version which performs CBJ and uses DBT $Nogoods$. To emphasize its advantage, the results of the most successful versions of MAC and FC are presented in Figure 8. In the case of structured problems the largest difference in performance is for tight problems.

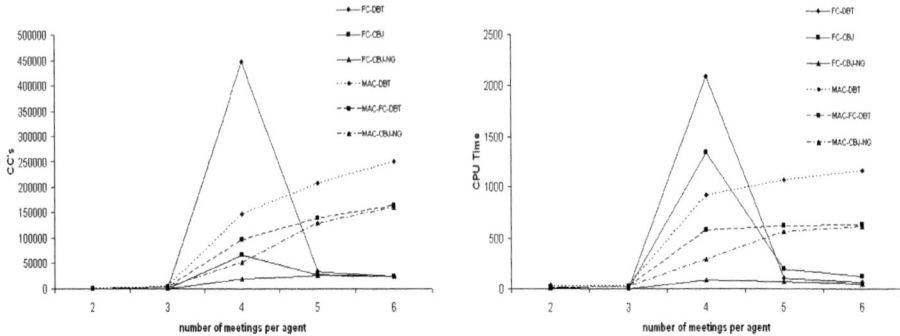

Fig. 7. Constraints checks and CPU-time performed by the different algorithms on Random Meeting Scheduling Problems

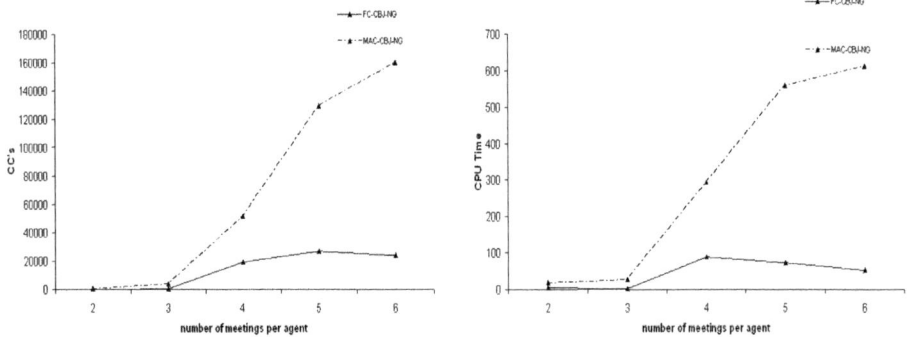

Fig. 8. Constraints checks and CPU-time performed by the best versions of MAC and FC on Random Meeting Scheduling Problems

7   Discussion

Dynamic Backtracking (DBT) was proposed by [Gin93] as a mechanism that enables the search algorithm to perform backjumping while preserving the assignments of variables which were jumped over. The proposed algorithm was found to *harm* the effect of ordering heuristics and to perform poorly compared to standard conflict directed backjumping when powerfull ordering heuristics are used [Bak94].

Enhancing DBT with local consistency methods (FC, MAC) was proposed by [JDB00] who found that the most successful version is *MAC-DBT*. This version was also found to outperform standard versions of MAC[MOSHE PLEASE ADD THE REFERENCE].

The results presented in the present paper demonstrate that the advantage of *MAC-DBT* over *FC-DBT* exists *only when static order is maintained*. When dynamic ordering heuristic is used, *FC-DBT* runs faster than *MAC-DBT*. Our presented results, both theoretical and empirical, show clearly the advantage of performing a combined version of *MAC-DBT* with explicit froward-checking, over the *MAC-DBT* of [JDB00].

The best performing algorithms presented in this paper do not preserve assignments which were jumped over (as in [Gin93]). As a result, the properties of the ordering heuristic are preserved in contrast to standard DBT. Both versions of DBT with lookahead (FC and MAC) benefit from using the DBT mechanism for storing $Nogood$ explanations. The benefit arises from the ability to determine which value should be restored to a variable's domain. Updated domains during backjumping enhance ordering heuristics that are based on domain size. The algorithms proposed by the present paper FC-CBJ-NG and MAC-CBJ-NG, improve on both MAC-DBT and FC-DBT. The best performing algorithm was found to be FC-CBJ-NG. Its advantage over all other versions is most pronounced on structured (*Meeting Scheduling*) problems.

References

[Bak94]   Baker, A.B.: The hazards of fancy backtracking. In: Proceedings of the 12th National Conference on Artificial Intelligence (AAAI 1994), Seattle, WA, USA, July 31 - August 4, vol. 1, pp. 288–293. AAAI Press, Menlo Park (1994)

[BFR95]   Bessière, C., Freuder, E., Régin, J.: Using inference to reduce arc consistency computation. In: IJCAI 1995, pp. 592–598 (1995)

[BR96]   Bessière, C., Regin, J.C.: Mac and combined heuristics: two reasons to forsake fc (and cbj?) on hard problems. In: Freuder, E.C. (ed.) CP 1996. LNCS, vol. 1118, pp. 61–75. Springer, Heidelberg (1996)

[CvB01]   Chen, X., van Beek, P.: Conflict-directed backjumping revisited. Journal of Artificial Intelligence Research (JAIR) 14, 53–81 (2001)

[Dec03]   Dechter, R.: Constraint Processing. Morgan Kaufman, San Francisco (2003)

[DF02]   Dechter, R., Frost, D.: Backjump-based backtracking for constraint satisfaction problems. Artificial Intelligence 136(2), 147–188 (2002)

[Gin93]   Ginsberg, M.L.: Dynamic backtracking. J. of Artificial Intelligence Research 1, 25–46 (1993)

[GW99]   Gent, I.P., Walsh, T.: Csplib: a benchmark library for constraints. Technical report, Technical report APES-09-1999 (1999), http://csplib.cs.strath.ac.uk/; A shorter version appears in the Proceedings of the 5th International Conference on Principles and Practices of Constraint Programming (CP 1999)

[HE80]   Haralick, R.M., Elliott, G.L.: Increasing tree search efficiency for constraint satisfaction problems. Artificial Intelligence 14, 263–313 (1980)

[JDB00]   Jussien, N., Debruyne, R., Boizumault, P.: Maintaining arc-consistency within dynamic backtracking. In: Dechter, R. (ed.) CP 2000. LNCS, vol. 1894, pp. 249–261. Springer, Heidelberg (2000)

[KvB97]   Kondrak, G., van Beek, P.: A theoretical evaluation of selected backtracking algorithms. Artificial Intelligence 21, 365–387 (1997)

[MH86]   Mohr, R., Henderson, T.C.: Arc and path consistence revisited. Artif. Intell. 28(2), 225–233 (1986)

[Pro93]   Prosser, P.: Hybrid algorithms for the constraint satisfaction problem. Computational Intelligence 9, 268–299 (1993)

[Pro96]   Prosser, P.: An empirical study of phase transitions in binary constraint satisfaction problems. Artificial Intelligence 81, 81–109 (1996)

[Smi96]   Smith, B.M.: Locating the phase transition in binary constraint satisfaction problems. Artificial Intelligence 81, 155–181 (1996)

Author Index